W9-AFS-510

The New American Grandparent

THE NEW AMERICAN GRANDPARENT

A Place in the Family, A Life Apart

Andrew J. Cherlin

Frank F. Furstenberg, Jr.

Basic Books, Inc., Publishers

NEW YORK

Library of Congress Cataloging-in-Publication Data

Cherlin, Andrew J., 1948–
The new American grandparent.

Includes index.
1. Grandparents—United States. 2. Grandparent and child—United States. I. Furstenberg, Frank F., 1940– . II. Title.
HQ759.9.C44 1986 306.8'7 85-73884
ISBN 0-465-04993-1

Printed in the United States of America
Designed by Vincent Torre
86 87 88 89 HC 9 8 7 6 5 4 3 2 1

To The Memory of Our Grandparents

CONTENTS

ACKNOWLEDGMENTS

This study was supported by grant number AG02753 from the National Institute on Aging (NIA). The funds from NIA allowed us to conduct telephone interviews with 510 grandparents throughout the country whose children and grandchildren had been interviewed during an earlier study. We also were able to conduct follow-up interviews in the homes of some of the grandparents in our sample. The original focus of the study was on the effects of divorce on grandparent-grandchild relations, an emphasis that is apparent in chapter 6 and elsewhere throughout the book. But during the course of our research, we broadened the topic to include a more general examination of the role of grandparents in the contemporary family. Needless to say, we greatly appreciate the support of NIA, Dr. Matilda White Riley, Associate Director for Behavioral Sciences Research, and the project officer, Ms. Kathleen Bond.

We also benefited from the assistance and suggestions of many professional colleagues. Nicholas Zill and James Peterson of Child Trends, Inc., graciously provided access to the National Survey of Children, onto which our grandparent survey was grafted. They also provided valuable advice throughout the project. Ellin Spector of the Institute for Survey Research at Temple University was a fund of information and enthusiasm as she directed the field work for the survey. We presented an earlier version of the material in chapter 6 at the Wingspread Conference on Grandparenting and Family Connections, sponsored by the National Institute for the Family and the William Petschek National Jewish Family Center of the American Jewish Committee, at Racine, Wisconsin, on October 9–11, 1983. The discussions and presentations at that conference helped to stimulate our work. An earlier version of chapter 3 appeared in Vern L. Bengtson and Joan F. Robertson, editors, *Grandparenthood* (Beverly Hills, CA: Sage Publications, 1985). Professor Bengtson provided especially helpful comments on these early papers and later on the entire manuscript. We also wish to thank W. Andrew Achenbaum, Colleen Johnson, John Modell, Yehuda Rosenman, Daniel Scott Smith, Arland Thornton, and Lillian Troll for taking the time to read and comment on all or part of the manuscript. At Johns Hopkins, Su-Hoon Lee was a valuable research assistant and Shirley Sult ably typed and retyped the various drafts. At the University of Pennsylvania we are indebted to Laura Blakesley, Kathy Harris, and Mary Elizabeth Hughes, who provided research assistance, and to Julia Robinson, who helped in both the management of the project and the preparation of the manuscript. We also wish to acknowledge the assistance of staff members who set up our preliminary interviews at senior citizen centers in an eastern city, even though we cannot thank them or their centers by name for reasons of confidentiality. And of course we are grateful to the grandparents around the country who took the time to participate in our study.

Finally, we want to acknowledge the contribution of Cheryl Allyn Miller. Dr. Miller came to the Department of Sociology at Johns Hopkins as a postdoctoral fellow soon after this project began. She assisted in the design of the survey and conducted most of the follow-up interviews from which quotations are drawn in this book. Tragically, she became ill late in 1983 and died in 1984. We miss her as a colleague and, above all, as a friend.

The New American Grandparent

The New American Grandparent is factually accurate; however, all names and identifying characteristics of the respondents have been changed.

1

Introduction

On a summer day in 1982, we began our study of American grandparents by driving to a senior citizen center in an eastern city. The neighborhood had once been solidly Jewish. But since the 1960s many of the Jews had been moving to the suburbs, and other ethnic groups had taken their place. As in other neighborhoods that have experienced this familiar process of urban succession, older people, limited economically and reluctant to leave their longtime homes and friends, were the last to leave. At the senior citizen center we found a group of about a dozen Jewish residents of the neighborhood who had volunteered to talk with us about being grandparents.

The idea of studying the nature of grandparenthood had intrigued us because of the peculiarly marginal nature of American grandparents. Until recently, they were a mere footnote in the social scientific literature on the family. Most of the articles on the subject prior to about 1960 considered grandparents to be isolated, cut off from kin, a social problem. Psychoanalysts published studies in the 1950s with titles such as "The Role of Grandparents in the Psychology of Neurosis" and "The Grandparent Syndrome," about the potential harm that their meddlesome behavior could bring to children and grandchildren.[1] Writing at a time when divorce rates were lower and birth rates were higher, sociologists in the 1940s and 1950s focused on the nuclear family, consisting of mother, father, and children. They observed that marital bonds increasingly were based on sentiment rather than obligation, but they failed to notice that the same trends were affecting ties to kin outside the nuclear family. Concerned largely with marriage, they simply failed to examine other aspects of American kinship. By most accounts, being a grandparent was an unimportant role that merited little attention.[2]

And yet lately there has been something of a rediscovery of grandparents. As divorce and remarriage increased in the 1960s and 1970s, the nuclear family receded in importance, giving way to a variety of family forms. Social scientists became more receptive to exploring the complex network of kinship ties that link individuals across households and from generation to generation. Some began to look at the role of grandparents. A psychologist argued that grandparents provide a valuable service as the "family watchdogs," standing ready to step in with assistance in times of need.[3] A child psychiatrist argued in a recent book that more grandparents could—and should—play a vital role in enriching their grandchildren's lives.[4]

Moreover, grandparenthood began to receive more attention from the general public. Popular television series such as *The Waltons* conjured up a nostalgic picture of large, extended families with strong, loving grandparents. In the late 1970s and early 1980s most states enacted so-called "grandparents' rights" legislation, giving grandparents who are prevented from seeing their grandchildren by the custodial parent the legal standing to request visitation privileges from the court. And in what may be the ultimate symbol of political and commercial recognition, grandparents now have their own day. In 1978, Congress passed legislation setting the first Sunday after Labor Day as Grandparents Day. The establishment of the holiday culminated a six-year campaign organized by a retired Atlanta real estate developer, Mike Goldgar, and a network of like-minded citizens who successfully lobbied members of Congress. Once the holiday was established, it was quickly supported by the greeting card and florist industries. Grandchildren and their parents now purchase about four million Grandparents Day cards each year.

To be sure, much of this attention reflects the growing number of older people, and hence grandparents, in the population (though not all grandparents are old). State legislators who worked for grandparents' rights undoubtedly were motivated by the increasing proportion of older voters. When Grandparents Day legislation reached the House floor in 1978, only eight representatives opposed it. Voting against grandparents is political suicide. The renewed attention to grandparents also comes at a time when cost-conscious policy makers are searching for ways to substitute family support for government assistance programs. In 1985, for example, the state of Wisconsin enacted a law that requires parents of teenaged sons and daughters to support any grandchildren until the unwed teenaged parents reach age eighteen. Moreover, the popularity of the extended family image may reflect the anxiety many Americans feel about recent changes in family life, such as the increase in divorce and the fall in family size.

Yet we still do not know much about grandparents today—about what it means to be a grandparent, about the roles grandparents play in family life, or about the ways in which grandparenthood may have changed since today's grandparents were grandchildren. We thought these questions worth pursuing in some detail. The very marginality of grandparents should mean that their role is readily influenced by broad social trends. It seemed to us that recent changes in grandparenthood could provide an important case study of the more general changes that have occurred in family and personal relationships. Finding out more about grandparents might provide valuable insights about the strengths, weaknesses, and dilemmas of the contemporary American family.

And so we found ourselves at the senior citizen center. The three of us—Cherlin, Furstenberg, and our research associate, Cheryl Miller—discussed how to conduct the group interview. How could we ask strangers about something as intimate as what it meant to be a grandparent? Would they open up to us? Would they have much to say beyond platitudes such as "We love them all?" We decided our best bet was to tell them a little about the objectives of our project and then throw ourselves upon their mercy. And that was what we did.

We parked our car in front of a converted store and were led past a noisy sing-along into a small activity room where our informants, sitting in a circle, awaited our arrival with curiosity. The first person to speak was an erect but frail-looking man who spoke slowly and deliberately:

> Mr. Stein is my name. Thank goodness we don't have any divorces in our family. We have two sons. One son has two daughters and one son; the other son's got three girls. Thank goodness we have what we call a lot of *naches* [pleasure] from the grandchildren. Our youngest granddaughter was just *bas mitvahed.* And for a *zehdee* [grandfather] to sit in the audience [voice breaking with emotion] and see his little girl get up there and make a speech. . . . It does the grandparents a lot of good. . . . And thank goodness we don't have any trouble with either one of our sons. That's all I gotta say . . . [breaks down crying]. I'm alright.

At first we thought that Mr. Stein's breakdown had less to do with being a grandparent than with his obviously poor health. After he had regained his composure, the person next to him in the circle started to speak. But when she also broke down while telling her story, we knew we had touched upon a role that elicited deep and powerful feelings, that whatever else might be said about being an older parent or a grandparent, it was a very meaningful part of these people's lives:

I have three children and they were married many, many years ago. And I remember my sister-in-law said to me, "Oh, you're so lucky your children are married, they're not in this crazy age where they're married, they're divorced," and before I knew it, bang! . . . Before I knew it, my son, after eighteen years, got divorced. And my younger daughter—she was married about ten or twelve years—she got divorced. And . . . [unable to continue].

Both the pleasure and the pain of being a grandparent were so deeply felt that they were difficult to talk about. But as we went around the circle, the grandparents did talk openly about their roles, introducing us to many themes we would hear again and again during our study.

INTERVIEWER: *What's the difference in the way things are between you and your grandchildren compared to your relationship, if you had one, with your grandparents?*

My grandmother was a matriarch. There's no comparison whatsoever between the grandmothers of the previous days and now. Our grandmother was the queen of the family, we looked up to her. She wasn't mean—she was good, she was kind—but she was the boss of a family. I am not a boss.

INTERVIEWER: *What's the difference between then and now?*

The difference is that in those days—it isn't respect, I don't know what word to call it, it wasn't fear, it wasn't respect, it was what we were taught. The grandmother is the queen of the family. You must never criticize her, even when she's wrong. I never dared open my mouth. [Pauses; starts again emphatically:] I couldn't be a companion. The only thing I didn't like is there was no companionship. There was love, an awful lot of love and respect, whereas with my grandchildren, it is companionship. I'm their pal; [turning to her husband] he's their pal. He's "Dan" and I'm "Ruth." Half of the time they don't say "Grandma," they say "Ruth" to me. I

> don't get angry. I wouldn't dare have thought of calling my grandmother by her name. . . . To be a grandmother and a grandfather of the present day cannot compare to being grandparents of other days. I think these days are much happier.

"I'm their pal." A different style of grandparenting had emerged during these people's lifetimes, a companionate style we would learn much more about. The reasons for its emergence included smaller families:

> My mother was one of nine children, and I was the eldest grandchild. And I was one of twenty grandchildren to my grandmother. So there were so *many.* Well, today we have a limited number, and every one counts. So there's closeness.

Another reason is a higher standard of living—more money to spread among fewer children and grandchildren:

> The financial thing is different now. Years ago there were a lot of children—parents had five, six, seven, eight children. They didn't have enough money to go around, and then if the children got married you had grandchildren. So what did you have to give them? Today there's two, three, if you have four it's a lot. So you can do for them what wasn't done for you. What you wanted done when you were a grandchild.

There is also more leisure time to spend with the grandchildren:

> I'm so happy with my grandchildren, which I wasn't with my children. What I mean is I was in business all the time. When they used to come to me and ask me a question, "Dad, why don't you do this with me?" [I'd reply] "I don't have time, I gotta tend to the business." I really tried my best to

give them all I could, but it wasn't enough. So I've made up with my grandchildren today what I lost from my children.

Another reason remained unspoken but was evident to all. They had lived long enough to become grandparents and were healthy enough to enjoy it, unlike many of their grandparents, who died young or were in poor health in their later years. In all, they were telling us, they felt closer to their grandchildren because they had more time, more energy, and more to give than their grandparents had; and they had fewer grandchildren to give to.

Yet hadn't the grandparent role become more limited as it became more companionate?

INTERVIEWER: *It sounds like some of you feel as though with that closeness you've lost a little of the authority that your grandparents had. Is that right?*

That's right, yeah.

And replied another grandparent:

Well, I hear most grandparents don't want to be bothered too much with authority. They still feel like they're young enough to go out and do what they couldn't do when they were young.

Many grandparents may indeed be relieved to surrender authority over their grandchildren to their grown children. Other grandparents would tell us how nice it is not to be responsible. But even those who wanted to take on more authority apparently found it difficult to do so. True to the high value Americans place on the autonomy of kin, the grandparents in the circle were loath to "interfere":

You gotta cut the apron strings.

Another replied:

> No, it's not cut the apron strings. I don't interfere in [her son and daughter-in-law's] lives. We never interfere. Look, my son lives in [a nearby town]. And if he doesn't call up and say, "Mom, would you like to come over," we maybe don't go over there for two years. Only for the holidays.

Over and over again, we would hear that good grandparents should not "interfere" in their children's lives, especially with regard to how the children were raising the grandchildren. It would turn out, however, that a few grandparents could cross this sensitive boundary and play an authoritative role in their grandchildren's upbringing. Finding out who they were and how they managed to breach the norm of noninterference would provide us with some important clues about grandparenthood and, more generally, the rights and obligations of kinship in America.

It also became clear from the group's comments that grandparenthood was like a career. It has a distinctive beginning period when the grandchildren are young:

> When they were small, we used to babysit, and we played with them, and we carried on. Today you don't do that.

In the later stages of the grandparental career, "you don't do that" in part because your children do not need your services as much. So you must work to keep your career active when the grandchildren are older:

> In order to get close you have to give *yourself.* Like, you call 'em up—"How about going out tonight?" or "How about me coming over to your house?" You know, the married ones. You got to give yourself in order to get something back.

INTERVIEWER: *And you didn't have to do that when they were young?*

No. When they were young, they depended more on you.

Even as they gave of themselves, the grandparents had to let go—much as parents do—when the grandchildren approached adulthood:

I love my grandchildren [ages fifteen and eighteen] and I'm sure they love me, but I rarely see them socially, other than Friday night dinner. . . . I don't go out with them.

INTERVIEWER: *So you're saying that as they get a little bit older. . . .*

They get away from you. They love to be with their peers.

Nevertheless, a grandparent, as we would learn later, can have a number of strategies for maintaining and prolonging a successful career.

We wondered whether this was an unusual group. Were the sense of closeness and the strong feelings they expressed typical of grandparents? Or had those who had the closest relationships and the most to say—the professional grandparents—volunteered to talk to us? Toward the end of the interview one of us asked whether the grandparents at the sing-along in the next room were any different.

INTERVIEWER: *Is this a special group?*

Well, there are the hidden ones. They don't have anything to say; if they do, they'll say, "Everything is good." But it's no good.

INTERVIEWER: *Wait. I'm a little confused about this. Are you telling me that out there there are a lot of people who don't have such good relationships?*

Several grandparents speaking at once replied:

> Yes. . . . Oh, yes. . . . *Tsaurus* [trouble]. . . . Sure, plenty of them.

But we were not able to get a clear explanation of what made this group so different from the "hidden ones," and we left the center with some skepticism about their claims to distinctiveness.

This skepticism increased the following day, when we conducted a second group interview at a senior citizen center in a predominantly Polish-American neighborhood. We heard many of the same themes:

> INTERVIEWER: *Did your grandmother behave the way grandparents behave today?*
>
> No, well, she was quiet.

Another person interrupts:

> You mean she didn't get on the couch and play cowboys like we did [laughter]?

The very thought of the old-time grandmother horsing around with the grandchildren was enough to set everyone laughing. Once again we heard that companionship is more common now:

> In our time we were more disciplined. We'd sit there. It's different today; the children are more open.

And it was said to be more common for the same reasons we had heard the day before, such as more leisure time:

> You have more time to play with them now. The parents are busy, but the grandparents have time and they can play with them now.

And advances in communication that had been taken for granted by a naive interviewer:

> I think you have a lot of feeling today. . . . The grandchildren will say, "I'll call you when we get home, Grandma, to let you know we got home safe." And then he'll call—"We're home and OK"—and hang up.
> INTERVIEWER: *Is there more of that than there used to be in the old days?*

Another person, interrupting:

> In the old days, we didn't have the telephone [laughter]!

The same constraints on grandparents also were apparent:

> INTERVIEWER: *Is it important to keep a good relationship with your daughters-in-law?*
> Very important.

Another replied:

> You smile even if it hurts.
> INTERVIEWER: *Sounds like it might be a little bit difficult.*
> Once in a while something will come up that I'll say she should have done differently. But I'll say they are the younger generation and I'm not gonna think about it, because if you think about it you only get hurt.

"You only get hurt" because there's nothing you can do about it without committing the gross impropriety of interfering in your children's lives. So you might as well not think about it. You might as well consider it to be something beyond your concerns, not your responsibility. The twin guidelines for successful grandparenting were as follows:

INTERVIEWER: *What makes a good grandparent?*

Love, companionship.

Another grandparent responds:

Respect your children's wishes.

But at this center we also saw that there might be "hidden ones," as someone had called them the previous day, grandparents for whom the role was too troubling to talk about. One woman, it became apparent, had come to our session seeking help. With great difficulty, she told her story:

My name is Joyce Green. I have one son, three grandchildren. Well, my story had a happy beginning but a sad ending. They separated and my daughter-in-law doesn't allow the children to see me. . . . She won't allow them to call me or come to visit me or anything. . . . I'm not lying. I had a daughter-in-law, I used to go there every Saturday, my son used to pick me up eight o'clock, and I'd go there and I'd tell her, "Today is your day to do what you want." . . . I'd take the kids' shoes, take the laces out, wash the laces, polish the shoes, wash the clothes, put 'em out on the line. And the neighbors used to think I was *her* mother. And then she'd tell my son I'd come there on a Saturday and cause her trouble. She was jealous of me and my son. See, I only had him, and we were real close. And she was real jealous of me. So she told me that if I stayed away they'd get along better. But they didn't.

INTERVIEWER: *Now, does he see the kids at all?*

She won't let him see them too often. . . . Oh, it hurt me. I used to cry, and people said, "You're gonna make yourself sick." 'Cause I wanted them to come; they used to come to me. And she just said they're not allowed to see me. . . . Even when my granddaughter graduated, I was hurt. I think I cried

> all night because I wasn't even there. I can't get over the oldest one—she's got a car, she could come to see me. Every time that phone rings I think, "Oh, it's her." . . . What I want to know, is there any way if I'd go to somebody that the children could come to see me?

Unprepared to deal with such a request, we outlined the court procedures that were available to grandparents in many states and suggested she see a staff member at the center about pursuing them. Her story, her frankness in sharing her grief, had moved everyone. We would see elements of this story again when we talked to other grandparents whose sons had divorced and were no longer living with their children: how difficult it was for the grandparent to find ways to see the grandchildren if the son did not see them very often; how important it was for the grandparent to maintain a cordial relationship with the former daughter-in-law.

A month later, we learned more about the variations in grandparenthood when we met with a group of black grandparents at a senior citizen center in another neighborhood of the same city. Most of them had lived with grandchildren at least temporarily and had helped with childrearing. Many retained authority in their families in a way that the Jewish and Polish grandparents thought had passed from the scene:

> I was always named "sergeant"—"Here comes the sergeant." I loved them [her grandchildren]. I *did* for them, and gave to them, so that they had an education, so that they had a trade. I went to school regularly to check on them; they didn't know I was coming.

Here was a grandmother who did things as a matter of course that most grandparents would consider meddlesome. The grandchildren were her responsibility as well as her children's.

And yet, she was careful to disavow any intention to interfere in her children's lives:

> I had said that when the day came that my children were going to be married, this was their life. And I didn't want to be a mother-in-law about whom they could say, "If it hadn't been for my mother-in-law. . . ." Any assistance I can give or suggestions I can give, alright, but you can very quickly and very easily be a mother-in-law who has created dissension in the family.

Although her behavior differed from what we had seen before, she and the other black grandparents accepted the same ideology: that parents and their adult children ought to be autonomous except when help is needed. Nor was the perceived trend toward more companionship much different between the black and white grandparents.

> INTERVIEWER: *How different is it to be a grandparent these days from when you were growing up?*
>
> I'm more open with my grandchildren.

Another grandparent added that:

> Things are different today. Grandparents are more indulgent because they don't want to lose their grandchildren's love.

The black grandparents we spoke to seemed just as concerned about losing love and gaining companionship as the Jewish or Polish grandparents. But for many of them, a more functional, parentlike role was overlaid on the companionate role we had already seen.

The group interviews convinced us that grandparenthood was a deeply meaningful role. From the variety of experiences that were encountered, we began to form an impression of the nature of intergenerational relations today. We also began to realize that the dominant style of interaction had changed over the past few generations. In the following chapters, we will describe the attitudes and behaviors that increasingly characterize American grandparents. But we should state at the outset that there always have been grandparents who resemble the currently prevalent style. What is new about American grandparents is the greater proportion who fit the pattern we will discuss. Because of the social changes we will describe in chapter 2, far more grandparents today must try to balance their sometimes conflicting desires for a place in the family and a life apart.

But first we had to address an important question: how representative were these interviews? In order for us to talk with them, the grandparents had to meet several criteria (besides the obvious one of living in the same city): they had to be over sixty-five; they had to be the type of older people who spend a lot of time at a senior citizen center; and they had to be willing to talk to strangers about their family lives while their friends were listening. Moreover, for all our efforts we had only met with a small number of grandparents. So we wondered how many "hidden ones" existed for whom grandparenting was very different; how many grandparents there were whose experiences with divorce had been as painful as Mrs. Green's; whether younger grandparents acted and felt similarly; whether the differences we perceived between black and white grandparents were real; and so forth.

Questions such as these, which deal with the representativeness of the findings and with the distribution of experiences, necessarily plague all small-scale, informal studies. They can be resolved best by undertaking a larger study in which the subjects are selected at random from the population. This is the method of the large-scale sample survey, most commonly exemplified by the "opinion polls" found in newspapers and magazines. Because the respondents in the survey are selected randomly, they can be presumed to represent the larger population from which they are drawn. And because of the random selection process, one can make statements such as "50 percent of all Americans believe . . ." with some confidence. Moreover, if the survey is large enough, one can compare the young versus the old, blacks versus whites, and so forth.

Fortunately, we had an opportunity to undertake a sample survey of American grandparents. We did so by "piggybacking" our grandparent interviews on an existing set of interviews with parents and children. This existing project had begun in 1976, when social psychologist Nicholas Zill received a grant from the Foundation for Child Development to undertake a national survey of American children. Zill and his associate, James Peterson, supervised interviews with a randomly selected, nationally representative sample of children between the ages of seven and eleven (who will be referred to as the "study children") and with each child's primary caretaker—the mother, in most cases. The purpose of these interviews was to find out about the well-being of children—their health, happiness, and problems.[5] The 1976 interviews did not include any questions about grandparents. Family relations, especially across the generations, was not a major focus of the study.

In 1981, Furstenberg, Zill, and Peterson were awarded funds by the National Institute of Mental Health and the Foundation for Child Development to do a second study of some of the children. The topic that most concerned them was how children adjust when their parents divorce, so they reinterviewed all the children whose parents' marriages had disrupted or were experiencing serious difficulties by 1976, and a randomly selected subsample of children from maritally stable homes. The primary caretaker was also reinterviewed. Shortly before the 1981 interviews were carried out, Furstenberg and Cherlin discussed their mutual interest in learning more about American grandparents and their relationships with their children and grandchildren. They realized that if interviews with grandparents could be added to the project, the result would be a unique, three-generational national survey of families. Consequently, they inserted the following question into the 1981 parent interviews: "As part of our study, we would also like to conduct a telephone interview with [the child's] grandparents with your permission. May I have their names, addresses, and phone numbers?" The interviewer then wrote down the names, addresses, and phone numbers that were volunteered by the parent for the grandparents on her side, the grandparents on her spouse's side, and, if she was previously divorced, the grandparents on her ex-spouse's side.

Armed with the data from the child and parent interviews and with the telephone numbers of the children's grandparents, Cherlin and Furstenberg then applied to the National Institute on Aging for funds to carry out the grandparent interviews. In 1982, the Institute awarded us a grant. The group interviews in the summer of 1982 were our first attempts to learn more about grandparenthood in preparation for the national study. Throughout the summer and fall, we continued to interview grandparents while developing a questionnaire that could be used over the telephone. (The national survey had to be done over the telephone because the grandparents were spread all over the country, and face-to-face interviews would have been too costly.) Then, between February and April of 1983, professional interviewers at the Institute for Survey Research at Temple University, under our direction, conducted telephone interviews with 510 grandparents of the children in the 1981 national study.

Before the interviewing began, we were a bit concerned about whether telephone interviews would be adequate, fearing that some elderly grandparents might find them tiring or might be hard of hearing. But our fears proved groundless; our biggest problem was getting the grandparents off the phone. When interviewers called the grandparents and asked if they would like to talk about their grandchildren, most were only too pleased to oblige. Some grandparents, when asked a specific question, discoursed at great length about their families. We found it necessary to train the interviewers to keep the pace moving along because the interviews were limited, due to cost considerations, to forty minutes. Our interviewers rated 64 percent of the grandparents as "very interested or enthusiastic" subjects, 32 percent as "somewhat interested," and just 4 percent as "indifferent or reluctant." According to the interviewers, 92 percent were "attentive, involved, and responsive" during the interview. Although some people think that survey research is increasingly perceived as intrusive and burdensome, our respondents thought otherwise: they clearly enjoyed themselves. We talked with one grandmother afterward:

> INTERVIEWER: *Were there any reactions or feelings you had after the interview?*
>
> No. Well, I just thought that it was interesting because I had never participated in anything like that before. And I didn't think any of the questions were things that I didn't want to answer. They told me when they first started that if there was anything I objected to answering, I could just skip it, but there wasn't anything. I thought it was interesting, all of the different phases of life—your own children and the way you brought them up and the way they are, compared to the way your grandchildren are, and things like that. I thought it was very interesting.

Still, relatively brief, structured interviews can provide mainly statistical information. One gains representativeness but loses the richness of experience that is conveyed by longer, less formal conversations. In order to recapture some of the richness we found in our group interviews, we conducted follow-up interviews with thirty-six of the grandparents in the national study. These follow-up interviews were conducted in the grandparents' homes, using a much less structured set of questions. The thirty-six informants were a diverse group that included residents of Northeastern, Southern, Midwestern, and Western states. Quotations from these interviews, which were taped and transcribed, will be used throughout the book to illustrate and clarify the quantitative findings from the larger survey.

When combined with the interviews of parents and children, our study becomes what we believe is the first nationally representative three-generational survey. Yet our study does have some limitations, most of which result from obtaining the names of the grandparents from an existing sample of parents and children, rather than just going out and finding a sample of grandparents. This piggy-back procedure was necessary—it brought the cost of the interviews down to a level where we were able to obtain financial support—and also desirable, in that it allowed us to merge information from three generations in each family. But it meant that, rather than simply interviewing a nationally representative sample of grandparents, we were really interviewing the grandparents of a nationally representative sample of children. The implications of this rather subtle distinction are as follows.

First, the study children were almost all between the ages of thirteen and seventeen by 1983. Consequently, our survey presents information on grandparents who had teenaged grandchildren and who therefore were older, on average, than the typical American grandparent. Twenty-two percent of the grandparents in this survey were under sixty, 18 percent were sixty to sixty-four, 43 percent were sixty-five to seventy-four, and 17 percent were seventy-five or over. Much of our attention will be on the relationship between grandparents and teenaged grandchildren, although we will often discuss the grandparents' relationships with other grandchildren. Intergenerational ties may be weaker when the youngest generation is in adolescence, so it is possible that the relationships between these grandparents and their younger grandchildren were stronger.

Second, we had to rely on the parents to provide us with the names of the grandparents. Twenty-one percent of the parents refused to provide us with any names. We suspect that some of the refusals were prompted by the illnesses or infirmities of elderly grandparents. But we present evidence in appendix 1 that parents tended to withhold the names of the grandparents who lived farther away or who did not provide as much support. As we discuss in detail in appendix 1, these and other possible biases do not alter any of the major conclusions of our study, but they may make our portrait of grandparent-grandchild relations a bit too rosy. Throughout the book, we will note the points at which potential sampling biases may most effect our findings.

Another limitation arose from our decision to overrepresent grandmothers in the survey. We did so because most of the parent respondents were female and, according to many studies, women are more deeply involved in kin networks. Yet we obtained only sixty-four interviews with grandfathers instead of the one hundred we had designed our sampling procedures to produce. Grandfathers are older, on average, than grandmothers and more likely to be in poor health. The younger ones, moreover, are employed and therefore are harder to reach over the phone. But in addition, it seemed clear that grandfathers were less willing than grandmothers to talk about their grandchildren to an interviewer—thus confirming once again, it would appear to us, their lesser involvement with kin. It may be that the grandfathers we did manage to interview were more involved with their grandchildren than other grandfathers.

These and other technical issues are discussed further in appendix 1, which also includes the questionnaire. Despite these limitations and complexities, we would immodestly maintain that this study represents one of the better data sets about intergenerational relations ever assembled. In the ensuing chapters, we will draw upon the statistical evidence and the personal conversations we obtained in 1983 to pursue the themes and questions that arose from our preliminary interviews. First, however, let us step back and examine the historical evidence about the ways in which grandparenthood has changed in this century.

2

The Modernization of Grandparenthood

Writing a book about grandparents may seem an exercise in nostalgia, like writing about the family farm. We tend to associate grandparents with old-fashioned families—the rural, extended, multigenerational kind much celebrated in American mythology. Many think that grandparents have become less important as the nation has become more modern. According to this view, the shift to factory and office work meant that grandparents no longer could teach their children and grandchildren the skills needed to make a living; the fall in fertility and the rise in divorce weakened family ties; and the growth of social welfare programs meant that older people and their families were less dependent on each other for support. There is some truth to this perspective, but it ignores a powerful set of historical facts that suggest that grandparenthood—as a distinct and nearly universal stage of family life—is a post–World War II phenomenon.

Consider first the effect of falling rates of death. Much of the decline in mortality from the high preindustrial levels has occurred in this century. According to calculations by demographer Peter Uhlenberg, only about 37 percent of all males and 42 percent of all females born in 1870 survived to age sixty-five; but for those born in 1930 the comparable projections are 63 percent for males and 77 percent for females.[1] The greatest declines in adult mortality have occurred in the last few decades, especially for women. The average number of years that a forty-year-old white woman could expect to live increased by four between 1900 and 1940; but between 1940 and 1980 it increased by seven. For men the increases have been smaller, though still substantial: a two-year increase for forty-year-old whites between 1900 and 1940 and a four-year increase between 1940 and 1980. (The trends for nonwhites are similar.) Consequently, both men and women can expect to live much longer lives than was the case a few decades ago, and more and more women are outliving men. In 1980, the average forty-year-old white woman could expect to live to age eighty, whereas the average forty-year-old white man could expect to live only to age seventy-four. As a result, 60 percent of all the people sixty-five and over in the United States in 1980 were women.[2] Thus, there are many more grandparents around today than just a few decades ago simply because people are living longer—and a majority of them are grandmothers.

This decline in mortality has caused a profound change in the relationship between grandparents and grandchildren. For the first time in history, most adults live long enough to get to know most of their grandchildren, and most children have the opportunity to know most of their grandparents. A child born in 1900, according to Uhlenberg, had a better than nine-out-of-ten chance that two or more of his grandparents would be alive. But by the time the child reached age fifteen, the chances were only about one out of two that two or more of his grandparents would still be alive. Thus, some children were fortunate enough to establish relationships with grandparents, but in many other families the remaining grandparents must have died while the grandchild was quite young. Moreover, it was unusual for grandchildren at the turn of the century to know all their grandparents: only one in four children born in 1900 had four grandparents alive, and a mere one in fifty still had four grandparents alive by the time they were fifteen. In contrast, the typical fifteen-year-old in 1976 had a nearly nine-out-of-ten chance of having two or more grandparents still alive, a better than one-out-of-two chance of having three still alive, and a one-out-of-six chance of having all four still alive.[3] Currently, then, nearly all grandchildren have an extended relationship with two or more grandparents, and substantial minorities have the opportunity for extended relationships with three or even all four.

Indeed, Americans take survival to the grandparental years pretty much for granted. The grandparents we spoke to rarely mentioned longer life when discussing the changes since they were children. *Of course* they were still alive and reasonably healthy; that went without saying. But this taken-for-grantedness is a new phenomenon; before World War II early death was a much greater threat, and far fewer people lived long enough to watch their grandchildren grow up.

Most people are in their forties or fifties when they first become grandparents. Some observers have mistakenly taken this as an indication that grandparents are younger today than in the past. According to one respected textbook:

> Grandparenting has become a phenomenon of middle age rather than old age. Earlier marriage, earlier childbirth, and longer life expectancy are producing grandparents in their forties.[4]

But since the end of the nineteenth century (the earliest period for which we have reliable statistics) there has been little change in the average age at marriage. The only exception was the 1950s, when ages at marriage and first birth did decline markedly but only temporarily.[5] With the exception of the unusual 1950s, then, it is likely that the age when people become grandparents has stayed relatively constant over the past century. What has changed is the amount of time a person spends as a grandparent: increases in adult life expectancy mean that grandparenthood extends into old age much more often. In our national sample of the grandparents of teenagers, six out of ten had become grandparents while in their forties. When we interviewed them, however, their average age was sixty-six. Grandparenting has been a phenomenon of middle age for at least the past one hundred years. The difference today is that it is now a phenomenon of middle age *and* old age for a greater proportion of the population. To be sure, our notions of what constitutes old age also may have changed, as one woman in our study implied when discussing her grandmother:

> She stayed home more, you know. And I get out into everything I can. That's the difference. That is, I think I'm younger than she was at my age.

Moreover, earlier in the century some middle-aged women may have been too busy raising the last of their own children to think of themselves as grandmothers. Nevertheless, in biological terms, the average grandparent alive today is older, not younger, than the average grandparent at the turn of the century.

Consider also the effects of falling birth rates on grandparenthood. As recently as the late 1800s, American women gave birth to more than four children, on average.[6] Many parents still were raising their younger children after their older children had left home and married. Under these conditions, being a grandparent often overlapped with being a parent. One would imagine that grandparenthood took a back seat to the day-to-day tasks of raising the children who were still at home. Today, in contrast, the birth rate is much lower; and parents are much more likely to be finished raising their children before any of their grandchildren are born. In 1900, about half of all fifty-year-old women still had children under eighteen; but by 1980 the proportion had dropped to one-fourth.[7] When a person becomes a grandparent now, there are fewer family roles competing for his or her time and attention. Grandparenthood is more of a separate stage of family life, unfettered by child care obligations—one that carries its own distinct identification. It was not always so.

The fall of fertility and the rise of life expectancy have thus greatly increased the supply of older persons for whom grandparenthood is a primary intergenerational role. To be sure, there always have been enough grandparents alive so that everyone in American society (and nearly all other societies, for that matter) was familiar with the role. But until quite recently, an individual faced a considerable risk of dying before, or soon after, becoming a grandparent. And even if one was fortunate enough to become a grandparent, lingering parental obligations often took precedence. In past times, when birth and death rates were high, grandparents were in relatively short supply. Today, as any number of impatient older parents will attest, grandchildren are in short supply. Census data bear this out: in 1900 there were only twenty-seven persons aged fifty-five and over for every one hundred children fourteen and under; but by 1984 the ratio had risen to nearly one-to-one. In fact, the Bureau of the Census projects that by the year 2000, for the first time in our nation's history, there will be more persons aged fifty-five and over than children fourteen and under.[8]

Moreover, technological advances in travel and long-distance communication have made it easier for grandparents and grandchildren to see or talk to each other. As mentioned in chapter 1, the grandparents at one senior citizen center had to remind us that there was a time within their memories when telephone service was not universal. We tend to forget that only fifty years ago the *Literary Digest* predicted a Landon victory over Roosevelt on the basis of responses from people listed in telephone directories—ignoring the crucial fact that telephones were to be found disproportionately in wealthier, and therefore more often Republican, homes. As late as the end of World War II, only half the homes in the United States had a telephone. The proportion rose quickly to two-thirds by the early 1950s and three-fourths by the late 1950s.[9] Today, more than 97 percent of all homes have telephones.[10] About one-third of the grandparents in our survey reported that they had spoken to the study child on the telephone once a week or more during the previous year.

Nor did most families own automobiles until after World War II, as several grandparents reminded us:

> I could be wrong, but I don't feel grandparents felt as close to grandchildren during that time as they do now. . . . Really back there, let's say during the twenties, transportation was not as good, so many people did not have cars. Fortunately, I can say that as far back as I remember my father always had a car, but there were many other people who did not. They traveled by horse and buggy and some even by wagons. And going a distance, it did take quite some time. . . .

Only about half of all families owned automobiles at the end of the war.[11] Even if a family owned an automobile, long trips still could take quite some time:

> Well, I didn't see my grandmother that often. They just lived one hundred miles from us, but back then one hundred miles was like four hundred now, it's the truth. It just seemed like clear across the country. It'd take us five hours to get there, it's the truth. It was an all-day trip.

But in the 1950s, the Federal government began to construct the interstate highway system, which cut distances and increased the speed of travel. The total number of miles driven by passenger vehicles increased from about 200 million miles in the mid-1930s to about 500 million miles in the mid-1950s to over a billion miles in the 1980s.[12] Not all of this increase represents trips to Grandma's house, of course; but with more cars and better highways, it became much easier to visit relatives in the next county or state.

But weren't grandparents and grandchildren more likely to be living in the same household at the turn of the century? After all, we do have a nostalgic image of the three-generation family of the past, sharing a household and solving their problems together. Surprisingly, the difference between then and now is much less than this image would lead us to believe. To be sure, there has been a drastic decline since 1900 in the proportion of older persons who live with their adult children. In 1900 the proportion was more than three out of five, according to historian Daniel Scott Smith; in 1962 it was one out of four; and by 1975 it had dropped to one in seven. What has occurred is a great increase in the proportion of older people who live alone or only with their spouses. Yet the high rates of co-residence in 1900 do not imply that most grandparents were living with their grandchildren—much less that most grandchildren were living with their grandparents. As Smith's data show, older persons who were married tended to live with unmarried children only; children usually moved out when they married. It was mainly widows unable to maintain their own households who moved in with married children. Consequently, according to Smith's estimates, only about three in ten persons sixty-five and over in 1900 lived with a grandchild, despite the great amount of co-residence between older parents and their adult children. What is more, because of the relative shortage of grandparents, an even lower percentage of grandchildren lived with their grandparents. Smith estimates that about one in six children under age ten in 1900 lived in the same household with someone aged fifty-five or over.[13] Even this figure overestimates the number of children living with their grandparents, because some of these elderly residents were more distant kin, boarders, or servants.

There were just too many grandchildren and too few grandparents for co-residence to be more common. In the absence of more detailed analyses of historical censuses, however, the exact amount of change since 1900 canot be assessed. Nor was our study designed to provide precise estimates of changes in co-residence. But it is still worth noting that just 30 percent of the grandparents in our sample reported that at least one of their grandparents ever lived with them while they were growing up. And 19 percent reported that the teenaged grandchild in the study had lived with them for at least three months. Undoubtedly, some of the grandparents in our study had shared a household with some of their other grandchildren, although we unfortunately did not obtain this information. Thus, although our study provides only imperfect and incomplete data on this topic, the responses are consistent with our claim that the change in the proportion of grandparents and grandchildren who share a household has been more modest than the change in the proportion of elderly persons who share a household with an adult child.

Grandparents also have more leisure time today, although the trend is more pronounced for men than for women. The average male can now expect to spend fifteen years of his adult life out of the labor force, most of it during retirement. (The labor force comprises all persons who are working for pay or looking for work.) The comparable expected time was ten years in 1970, seven years in 1940, and only four years in 1900.[14] Clearly, a long retirement was rare early in this century and still relatively rare just before World War II. But since the 1960s, workers have begun to leave the labor force at younger ages. In 1961, Congress lowered the age of eligibility for Social Security benefits from sixty-five to sixty-two. Now more than half of all persons applying for Social Security benefits are under sixty-five.[15] Granted, some of the early retirees are suffering from poor health, and other retirees may have difficulty adjusting to their new status. Still, when earlier retirement is combined with a longer life span, the result is a greatly extended period during which one can, among other things, get to know and enjoy one's grandchildren.

The changes in leisure time for women are not as clear because women have always had lower levels of labor force participation than men. To be sure, women workers also are retiring earlier and, as has been noted, living much longer. And most women in their fifties and sixties are neither employed nor raising children. But young grandmothers are much more likely to be employed today than was the case a generation ago; they are also more likely to have aged parents to care for. Young working grandmothers, a growing minority, may have less time to devote to their grandchildren.

Most employed grandparents, however, work no more than forty hours per week. This, too, is a recent development. The forty-hour week did not become the norm in the United States until after World War II. At the turn of the century, production workers in manufacturing jobs worked an average of fifty hours per week. Average hours dropped below forty during the depression, rose above forty during the war, and then settled at forty after the war.[16] Moreover, at the turn of the century, 38 percent of the civilian labor force worked on farms, where long hours were commonplace. Even in 1940, about 17 percent of the civilian labor force worked on farms; but currently only about 3 percent work on farms.[17] So even if they are employed, grandparents have more leisure time during the work week than was the case a few decades ago.

They also have more money. Living standards have risen in general since World War II, and the rise has been sharpest for the elderly. As recently as 1960, older Americans were an economically deprived group; now they are on the verge of becoming an economically advantaged group. The reason is the Social Security system. Since the 1950s and 1960s, Congress has expanded Social Security coverage, so that by 1970 nearly all nongovernment workers, except those in nonprofit organizations, were covered. And since the 1960s, Congress has increased Social Security benefits far faster than the increase in the cost of living. As a result, the average monthly benefit (in constant 1980 dollars, adjusted for changes in consumer prices) rose from $167 in 1960, to $214 in 1970, to $297 in 1980.[18] Because of the broader coverage and higher benefits, the proportion of the elderly who are poor has plummeted. In 1959, 35 percent of persons sixty-five and over had incomes below the official poverty line, compared to 22 percent of the total population. By 1982 the disparity had disappeared: 15 percent of those sixty-five and over were poor, as were 15 percent of the total population.[19] The elderly no longer are disproportionately poor, although many of them have incomes not too far above the poverty line. Grandparents, then, have benefited from the general rise in economic welfare and, as they reach retirement, from the improvement in the economic welfare of the elderly.

Because of the postwar prosperity and the rise of social welfare institutions, older parents and their adult children are less dependent on each other economically. Family life in the early decades of the century was precarious; lower wages, the absence of social welfare programs, and crises of unemployment, illness, and death forced people to rely on their kin for support to a much greater extent than is true today. There were no welfare checks, unemployment compensation, food stamps, Medicare payments, Social Security benefits, or government loans to students. Often there was only one's family. Some older people provided assistance to their kin, such as finding a job for a relative, caring for the sick, or tending to the grandchildren while the parents worked. Sometimes grandparents, their children, and their grandchildren pooled their resources into a single family fund so that all could subsist. Exactly how common these three-generational economic units were we do not know; it would be a mistake to assume that all older adults were cooperating with their children and grandchildren at all times. In fact, studies of turn-of-the century working-class families suggest that widowed older men—past their peak earning capacity and unfamiliar with domestic tasks as they were—could be a burden to the households of their children, while older women—who could help out domestically—were a potential source of household assistance. Nevertheless, these historical accounts suggest that intensive intergenerational cooperation and assistance was more common that it is today.[20] Tamara Hareven, for example, studied the families of workers at the Amoskeag Mills in Manchester, New Hampshire, at the turn of the century. She found that the day-to-day cooperation of kin was necessary to secure a job at the mill, find housing, and accumulate enough money to get by.[21] Cooperation has declined because it is not needed as often: social welfare programs now provide services that only the family formerly provided; declining rates of illness, death, and unemployment have reduced the frequency of family crises; and the rising standard of living—particularly of the elderly—has reduced the need for financial assistance.

The structure of the Social Security system also has lessened the feelings of obligation older parents and their adult children have toward each other. Social Security is an income transfer system in which some of the earnings of workers are transferred to the elderly. But we have constructed a fiction about Social Security, a myth that the recipients are only drawing out money that they put into the fund earlier in their lives. This myth allows both the younger contributors and the older recipients to ignore the economic dependency of the latter. The elderly are free to believe that they are just receiving that to which they are entitled by virtue of their own hard work. The tenacity of this myth—it is only now breaking down under the tremendous payment burden of our older age structure—demonstrates its importance. It allows the elderly to accept financial assistance without compromising their independence, and it allows children to support their parents without either generation openly acknowledging as much.

All of these trends taken together—changes in mortality, fertility, transportation, communications, the work day, retirement, Social Security, and standards of living—have transformed grandparenthood from its pre–World War II state. More people are living long enough to become grandparents and to enjoy a lengthy period of life as grandparents. They can keep in touch more easily with their grandchildren; they have more time to devote to them; they have more money to spend on them; and they are less likely still to be raising their own children.

The Bonds of Sentiment

Hand in hand with these changes in the structure of grandparenthood have come great changes in the emotional content of the grandparent-grandchild relationship. Here we are pressing against the limits of historical scholarship; it is more difficult to document alterations in ways of thinking than in material conditions. But on the basis of historical studies and the reports of the grandparents in our study, it seems that there has been an increasing emphasis during this century on bonds of sentiment: love, affection, and companionship.

When we asked grandparents whether grandparenthood had changed since they were grandchilren, we heard stories of their childhood that differed somewhat from the idyllic image of the three-generational family of the past. Their grandparents, we were told, were respected, admired figures who often assisted other family members. But again and again, our informants talked about the emotional distance between themselves and their grandparents:

The only grandmother I remember is my father's mother, and she lived with us.

INTERVIEWER: *What was it like, having your grandmother live with you?*

Terrible [laughter]! She was old, she was strict. . . . We weren't allowed to sass her. I guess that was the whole trouble. No matter what she did to you, you had to take it. . . . She was good, though. She was real helpful. She used to do all the patching of the pants, and she was helpful. But, oh, she was strict. You weren't allowed to do anything, she'd tell on you right away.

INTERVIEWER: *So what difference do you think there is between being a grandparent when you were a grandchild and being a grandparent now?*

It's different. My grandma never gave us any love.

INTERVIEWER: *No?*

Nooo. My goodness, no, no. No, never took us anyplace, just sat there and yelled at you all the time.

INTERVIEWER: *Did you have a lot of respect for your grandmother?*

Oh, we had to—whether we wanted to or not, we had to.

INTERVIEWER: *Do you think your grandchildren have as much respect for you now as you had for your grandmother?*

I don't think so, no. Because I think if my grandchildren have something to say, they'll come up and say it. I mean, they won't hold it back. . . . Whereas before you couldn't even speak out.

Grandma may have helped out, and she certainly was respected, even loved; but she often was an emotionally distant figure.

Gunhild Hagestad analyzed all mentions of grandparents in two volumes of *Good Housekeeping* from the 1880s. Most of the items were poems, often describing an old grandmother who might be sitting quietly by the fire. Describing the subjects of such odes as "Grandma—God Bless Her" (1887), Hagestad writes: "Seldom, if ever, was the 1880s grandmother described as dealing with the nitty-gritty aspects of everyday family life. The frail figure by the fire was not *withdrawn* from everyday living, but was *above* it. She had a place on a pedestal, and she had earned it."[22]

Many of our informants were from immigrant families. In a study of Italian immigrant families in Buffalo, New York, Virginia Yans-McLaughlin noted that the elderly were honored and revered and that grandmothers who helped with household tasks and childrearing were highly valued family members.[23] But our informants also spoke of the ways in which emotional distance could be compounded by culture and language:

> Well, see, my grandmother was Polish, and she couldn't speak English. But she used to come every Sunday, and when she'd come, we thought the world of her. And she'd try to speak to us—it was hard. It was hard for her to speak to us, because we all spoke English and she was Polish.
>
> INTERVIEWER: *I see. Did you ever go to church with her?*
>
> No, no.
>
> INTERVIEWER: *Did you joke or kid at all with her?*
>
> No, we never did it.
>
> INTERVIEWER: *You never went shopping together?*
>
> No.
>
> INTERVIEWER: *Day trips . . .*
>
> No, no.
>
> INTERVIEWER: *Did she ever give you money?*
>
> Oh yes, she used to treat us, she'd give us a nickel or something, and we thought it was a whole lot she gave us.
>
> INTERVIEWER: *How about disciplining? Did she discipline you?*
>
> Well, like I said, it was hard for us to make contact with one another because she was from Poland and we were born in America.

In his well-known study of the black family, E. Franklin Frazier noted the prestige, importance, and dignity of the black grandmother, the "guardian of the generations," during and after slavery. Even during the great twentieth-century migration of blacks to northern cities, the black grandmother, he wrote, "has not ceased to watch over the destiny of the Negro families."[24] We will have more to say in later chapters about the continued authority of black grandmothers today. But here let us note that even when our black informants spoke warmly and lovingly of their grandparents, they often commented on the lack of free expression and understanding:

> I grew up on a plantation. [When my grandparents] were no longer able to work in the fields, then they lived with one of their children or the other. . . . And my grandmother was a good cook. I loved that, she could make that good homemade bread. And my grandfather played with us a lot, and he would tell stories—he was a good storyteller. Then he played with us a lot and I liked that an awful lot. So they weren't grandparents that were rude or anything, they just were lovable grandparents. When they were around we just had a lot of fun.
>
> INTERVIEWER: *Do you think you were more friendly with your grandparents than you are with Peter* [her grandchild]*?*
>
> I think I'm more friendly with Peter than I was with my grandparents. Because . . . children couldn't express themselves at that time, when I was a child, like children can now. [They could] express themselves to me; I couldn't express myself to them. They would say that was being sass or bad or whatever. . . . I think the grandparents nowadays get a better understanding of their grandchildren than the grandparents then because we couldn't talk back to them. Once in a while we could ask them if we could do something—may I do this, or something. But that was it, almost. And maybe we could

> talk about school or something, about our lessons or something. They did the telling, you know.

"I couldn't express myself." This inability to communicate, to gain an understanding of the other person's thoughts and feelings, to be emotionally "close," to be friends, characterized grandparent-grandchild relations two generations ago, according to the recollections of our informants.

This is not to say that affective bonds were absent from the relations between young and old. On the contrary, both affection and respect were present. For example, Jane Range and Maris Vinovskis, who analyzed the content of short fiction in a popular nineteenth-century magazine, found that 43 percent of the elderly characters received respect from the younger characters and 48 percent received affection.[25] The shift, however, has been in the balance between respect and affection, as the responses to several questions in our national survey suggest. We asked all respondents who knew at least one of their grandparents, "Are you and [the study child] more friendly, less friendly, or about the same as your grandparents were with you?" Forty-eight percent said "more friendly"; only 9 percent said "less friendly." Similarly, 55 percent said that their relationship with the study child was "closer" than their relationship with their grandparents; just 10 percent said "not as close." When we asked about respect, most grandparents thought there had been no change, although 22 percent said they were more respectful to their grandparents versus only 2 percent who said their grandchildren were more respectful toward them. On the issue of authority, the responses were mixed: 28 percent thought they had more authority over the study child than their grandparents had over them, 22 percent thought their grandparents had more authority over them, and 50 percent saw no change. Perhaps there was less agreement about changes in authority because relatively few grandparents, even at the turn of the century, had substantial authority over their grandchildren.

To be sure, it is hard to judge how accurate these recollections are of the situation two generations ago. Moreover, the extent to which people can freely express their sentiments may have increased over time. But the story that the grandparents consistently told us fits with the demographic, economic, and technological developments we have discussed. It is easier for today's grandparents to have a pleasurable, emotion-laden relationship with their grandchildren because they are more likely to live long enough to develop the relationship; because they are not still busy raising their own children; because they have more leisure time; because they can travel long distances more easily and communicate over the telephone; and because they have fewer grandchildren and more resources to devote to them.

Above all, they are more likely to be their grandchildren's companions because of the increasing economic independence of the generations. When intensive economic cooperation and assistance was more common, older people and their families were bound together by instrumental ties of the sort we recognize today mainly in the lower class. Under these circumstances, one's obligations to kin took precedence over one's feelings toward them. Michael Anderson, for example, studied the effects of Poor Law relief on the quality of intergenerational relations in nineteenth- and early twentieth-century Britain. Poor relief, which was locally administered, varied widely from area to area. In areas where administration was strict and the burden of supporting the elderly fell more heavily on their children,

> . . . tensions were frequently increased both by the need to contribute and by the pressures applied by the authorities, so that, while the economic functions of kinship rose in importance, affective functions frequently declined.[26]

But where relief was more generous, he argued, bonds of sentiment became more visible. He cited an observer of one such area who noted in 1909 that a "weak sense of responsibility exists side by side with much natural affection and concern for each other's health and well being between parents and children among farm labourers."[27] From evidence such as this, Anderson reasoned that the replacement of the locally administered system with a uniform, nationwide system of income maintenance for the elderly in the mid-twentieth century "has markedly decreased the tensions and conflicts in family relationships" and allowed families to better perform "affective functions" for their members.[28]

Anderson's study suggests that with the increasing economic independence of the generations in this century, bonds of obligation have declined relative to bonds of sentiment. The responses of the grandparents in our study are consistent with this proposition. Freed of economic dependence and removed from day-to-day control over resources, today's grandparents, like the Jewish grandmother we met in chapter 1, strive to be their grandchildren's pals rather than their bosses. As another grandmother at the same senior citizen center said:

> The grandchildren really do love their grandparents more today than ever. I'm positive.
>
> INTERVIEWER: *Why?*
>
> You know why? The social status is almost the same. They're interested in what you're interested in.

They share not only interests but also relative economic equality: neither is dependent on the other. Anthropologists have described many other societies in which grandparents tend to have informal, friendly relationships with their grandchildren; this is one instance of the so-called "joking relationship" between kin.[29] Such informal relationships between grandparents and grandchildren, however, are not universal. Dorian Apple Sweetzer showed that "friendly equality" between grandparents and grandchildren is much more common in societies in which grandparents lack authority over their grown children.[30] Ours is also such a society.

Nevertheless, we do not wish to argue that the changing material conditions of grandparents in this century—demographic, technological, and economic—were the "cause" of the change in how they relate to their grandchildren in any simple sense. An increasing emphasis on sentiment in all family relations had been under way in the United States for some time. Historian Carl Degler claimed that the "modern American family"—characterized by affection and mutual respect between marriage partners, the primacy of the homemaking and child-rearing role of the wife, increasing child-centeredness, and smaller size—emerged between the American Revolution and 1830.[31] For instance, he cites research by Smith suggesting that parental control over children's choice of spouses and timing of marriage declined during that period (and, presumably, less constrained, "romantic" marriages increased).[32]

Attitudes toward children continued to evolve in the late nineteenth and early twentieth centuries. Viviana Zelizer has documented a shift in the value of children between the 1870s and 1930s from an economically useful asset to a priceless, but economically useless, object of sentiment. For example:

> In 1896, the parents of a two-year-old child sued the Southern Railroad Company of Georgia for the wrongful death of their son. Despite claims that the boy performed valuable services for his parents—$2 worth per month, "going upon errands to neighbors . . . watching and amusing . . . younger child"—no recovery was allowed, except for minimum burial expenses. The court concluded that the child was "of such tender years as to be unable to have any earning

> capacity, and hence the defendant could not be held liable in damages."[33]

Both the parents and the court argued solely on the basis of the child's economic worth. By the 1920s, according to Zelizer, this strict economic approach to children's worth had been replaced by a view of children as precious, priceless objects of love and affection. Judging the value of children by their potential earning power came to be seen as degrading, even immoral. Ironically, noted Zelizer, even as compulsory schooling and child labor laws reduced the economic contribution of children, awards for the wrongful death of children increased. Today, the economically useless but emotionally priceless child has great value:

> In January 1979, when three-year-old William Kennerly died from a lethal dose of fluoride at a city dental clinic, the New York State Supreme Court jury awarded $750,000 to the boy's parents.[34]

Moreover, almost every observer of historical change in the relationship between American husbands and wives has noted the increasing emphasis upon affection between the spouses. Degler claimed that this emphasis on love and affection as the basis for marriage emerged in the late eighteenth century.[35] Perhaps the best-known statement of the shift came in a widely used sociology of the family textbook written in 1945 by Ernest W. Burgess and Harvey J. Locke:

> The basic thesis of this book is that the family has been in historical times in transition from an institution with family behavior controlled by the mores, public opinion, and law to a companionship with family behavior arising from the mutual affection and consensus of its members. The companionship form of family is not to be conceived as having al-

> ready been realized, but as emerging. . . . The permanence of marriage is more and more dependent upon the tenuous bonds of affection, temperamental compatibility, and mutual interests.[36]

As for the elderly, recent historical scholarship suggests shifts in public attitudes, although historians disagree about the timing of the initial changes. David Hackett Fischer contended that the status of older people began to decline between 1770 and 1820. W. Andrew Achenbaum, however, argued that attitudes about the status of the elderly began to change for the worse between the Civil War and World War I.[37] By the end of World War I, wrote Achenbaum, old age had emerged as a social problem, although political actions designed to help solve the problem did not begin until the 1920s. In fact, Achenbaum stated flatly that "the fundamental modernization of old age in America took place after 1920."[38]

Fischer linked the shift in the status of the elderly to a shift in sentiment. Prior to the American Revolution:

> Even as most (though not all) elderly people were apt to hold more power than they would possess in a later period, they were also apt to receive less affection, less love, less sympathy from those younger than themselves. The elderly were kept at an emotional distance by the young. If open hostility between the generations was not allowed, affection was not encouraged either.[39]

Conversely:

> In modern America, as the social and economic condition of the aged worsened, their psychic condition may have grown a little better, in one way at least. As elders lost their authority within the society, they gained something in return.

> Within the sphere of an individual family, ties of affection may have grown stronger as ties of obligation grew weak.[40]

One must be skeptical of Fischer's implicit claim that most elderly were powerful, venerated figures prior to the Revolution. Carol Haber has argued that this may have been true for older people who controlled resources or maintained powerful positions—landowners, ministers, community leaders, and the like—but for the landless, the childless, and the otherwise impoverished, "gray hair and wrinkles seemed reason for contempt instead of honor; their age alone was not deemed worthy of respect."[41] Moreover, we know that in recent decades the social and economic condition of the aged has improved, not worsened. Nevertheless, we can infer from the grandparents in our study that the increasing emphasis on affection detected by scholars such as Fischer and Anderson was still continuing during their lifetimes.

Family change in the United States, then, has been a long-term process. The still incomplete historical record (the subdiscipline of family history is only about twenty years old) suggests that change has been discontinuous, occurring in great bursts and then subsiding for a time. One great period of change, several scholars tell us, lasted from the Revolution to the 1820s. A second one occurred about fifty years later: Zelizer argued that attitudes toward children changed dramatically between the 1870s and the 1930s; and it is known that during the same period a great debate about the rising divorce rate and a corresponding shift in attitudes toward divorce occurred.[42] We would suggest that future scholars, looking back on the current era, may view the 1960s and 1970s as another great period of change.

Many of the scholars cited here have advanced cultural or ideological explanations for the change in family values: the spread of the ideas of the Enlightenment or of the French and American Revolutions, or the changing modes of thought during the Progressive Era. (Others, such as Burgess and Locke, referred to a vaguely defined change from "traditional" to "modern" societies.) Without doubt, the long-term cultural changes influenced the evolution of the grandparent-grandchild relationship. But it also seems clear that the great transformation of this relationship throughout American society could not have taken place without the profound alteration of the material circumstances of grandparents that has occurred over the past several decades. To be sure, this transformation probably was under way among middle-class families before the twentieth century began. But the demographic, technological, and economic trends that have accelerated since 1940 were a necessary, though not sufficient, condition for the nearly universal emergence of a relationship based overwhelmingly on sentiment. Although child mortality rates dropped early in the century (contributing to the change in attitudes toward children), adult mortality rates dropped fastest, as we have noted, after 1940. Although the rising wages for male workers in the early decades of the century allowed middle-class wives to concentrate on childrearing and made marriage more of an emotional refuge from the outside world, the relative economic status of the elderly improved dramatically only after 1960. The result of these developments is the spread of a style of grandparenting characterized by emotionally satisfying leisure-time activities, a lack of direct responsibility for raising the grandchildren, and irregular direct assistance. It is a style that has only spread beyond the middle class since World War II because the demographic, technological, and economic changes that support it were not complete (or in some cases had hardly begun) until then. We will label it the *companionate* style of grandparenting, and we will have much more to say about it in the next chapter.

Grandparenting: Yesterday and Today

The grandparents in our study told us clearly what has been gained since they were grandchildren: a greater sense of understanding between the generations; more companionship, emotional warmth, and closeness; and a stronger emphasis on love. But it is not as clear what has been lost. There remains a feeling among many observers that modernization has removed something of value from the grandparent-grandchild relationship. A recent book by child psychiatrist Arthur Kornhaber and journalist Kenneth L. Woodward bemoaned the decline of "vital connections" between grandparents and grandchildren. When Kornhaber and Woodward asked their nonrandom sample of grandparents, "Do you have a good relationship with your grandchildren?" 75 percent responded "all of the time," 15 percent responded "most of the time," and only 10 percent responded "some of the time." Yet the authors dismissed these responses; actually, they argued, the grandparents were fooling themselves:

> The respondents did not understand that time and proximity were the basic foundations for a "good relationship." They confused spending a few pleasant moments together for a true relationship. They thought that they had a "good" relationship but upon deeper probing many expressed feelings that in reality, this relationship was not what they really wanted, that it had no relevance for them, that it wasn't enough.[43]

What is a "true relationship"? To Kornhaber and Woodward it is predicated upon grandparents living with or very near their grandchildren and spending a great deal of time with them.

More important, it arises from the kind of intense commitment to family that presumably developed when family members had to cooperate economically on a daily basis to make ends meet. In fact, the authors asserted:

> These adverse economic conditions were ideal for the development of vital connections between grandparents and grandchildren. Their lives converged in the critical dimensions of time and place; survival alone demanded commitment to the intrafamily network as a community of kin.[44]

It follows from this line of reasoning that one of the costs of prosperity is the loss of instrumental intergenerational ties—the decline of grandparents as mentors, role models, and caretakers.

This argument deserves to be taken seriously, for it probably is true that regular, material contributions of grandparents have declined in importance. As the stability of family life increased between the turn of the century and the 1950s, the contributions of kin became less necessary to the family's subsistence. The more that family life is unstable and precarious, then, the greater the supportive role of grandparents and other kin. Until recently, at least, the twentieth-century trend has been toward greater stability and, hence, less intensive support by grandparents.

But the argument is overstated. First, anyone who would hark back to the extended family of a few generations ago must keep in mind the harsh reality of high rates of death. One-fourth of the grandparents in our study never knew a grandparent. Many others undoubtedly lost their grandparents before they were old enough to establish much of a relationship. Moreover, as noted previously, most grandchildren, even if they had living grandparents, did not reside with them. What would be considered small geographical distances today were at one time major barriers to frequent visits. It seems safe to conclude that the three-generation economic unit, although undoubtedly more common in the past than it is today, was not part of the lives of most grandchildren for more than a short period of their childhood. To be credible, the critique of contemporary grandparenthood must be stripped of its nostalgic, romanticized view of the past.

Even so, it is fair to ask whether contemporary grandparents do anything for their grandchildren that is important. A number of observers have described the grandparental role as "empty," "ambiguous," "tenuous," or even "roleless."[45] In contrast, others have claimed that grandparents serve as stabilizing and unifying forces in their children's and grandchildren's lives. Hagestad has argued that one of the key functions of grandparents is "an elusive being here, a comforting presence which is not easily captured with the language and tools of social science."[46] The "comforting presence" provides intergenerational continuity and, should it be needed, a source of support. Lillian Troll has described grandparents as a latent source of support—the "family watchdogs," ever on the lookout for trouble and ready to provide assistance if a family crisis occurs.[47] These symbolic meanings of grandparenthood probably always have been with us. But the relative weight given to bonds of sentiment today makes them more central. For many grandparents and grandchildren today—especially those whose families are geographically dispersed, maritally intact, or relatively affluent—the function of intergenerational relationships may be defined largely in this symbolic way, by the elusive, but still deeply meaningful, being here.

But there are crosscurrents that suggest a continuing, even increasing, instrumental role for some grandparents. In low-income and minority families, for instance, parents and children sometimes rely heavily on grandparents and other kin. One study of low-income black families showed that grandmothers often provide crucial assistance to unmarried daughters who are raising children.[48] Middle-aged black men, according to a national study, were four times as likely as white men to have lived with their grandchildren for at least one year during the period 1966–1976.[49] Another study compared black and white middle-class families and found that black extended kin played a larger supportive role than white extended kin.[50] As we will report in later chapters, black grandparents still are deeply involved in the rearing of their grandchildren. Even among white middle-class families, older parents often contribute to the cost of buying a house or paying college tuition. The growing affluence of the elderly may increase their importance as financial resources.

In addition, the great rise in divorce during the 1960s and 1970s has made families less stable than they were in the 1950s. Between the early 1960s and the mid-1970s, the divorce rate doubled. If current rates continue, about one out of two recent marriages will end in divorce.[51] This high level of divorce might mean that many single parents will be turning to their own parents for support. Our survey, as we noted in chapter 1, included a large number of grandparents whose children had divorced after the birth of grandchildren. One of the major themes of this book will be the way in which the increase in divorce has altered the roles of grandparents once again—returning to some of them the greater functional role that was more widespread a few generations ago.

Regardless of what grandparents do for their grandchildren, there is the question of whether being a grandparent has a major impact on the grandparents themselves. Every contemporary observer of grandparents has noted how important the role is in symbolic terms to the grandparents. Becoming a grandparent is a deeply meaningful event in a person's life. Seeing the birth of grandchildren can give a person a great sense of the completion of being, of immortality through the chain of generations. It is an affirmation of the value of one's life and, at the same time, a hedge against death. Grandchildren are also a great source of personal pleasure. Freed from the responsibilities of parenthood, grandparents can unabashedly enjoy their grandchildren. In our exploratory interviews, two grandparents independently recited the same aphorism: "Your children are your principal; your grandchildren are your interest."

In order to advance our understanding of the nature and importance of contemporary grandparenthood, we need to take a detailed look at the relationship between grandparents and their grandchildren. As we argued in chapter 1, social scientists interested in the family have not paid much attention to grandparents. Perhaps this oversight is a result of the mistaken impression that grandparents belong to a bygone era. We hope this chapter has dispelled that impression. In the following chapters, we will present an account of the relationships between the grandparents and grandchildren in our national study. To repeat, we believe that a closer look at the changing nature of grandparenthood can help us better understand the broad, recent changes in American kinship and family life. Without a clearer picture of the strengths and limitations of intergenerational relations, our sense of the contemporary family is incomplete. Let us turn, then, to the reports of the American grandparents we interviewed.

3

Styles of Grandparenting

Many grandparents with whom we spoke had become specialists in recreational caregiving. They described themselves as playful companions, and the givers and receivers of love and affection. They told of an easygoing, friendly style of interaction with their grandchildren. When asked what they did with their grandchildren, they repeatedly mentioned emotionally satisfying, leisure-time activities. This is the essence of the companionate relationship, the dominant style of grandparenting among those with whom we talked. It was not, however, a universal style.

In our preliminary interviews and in our national survey, we found some grandparents and grandchildren who saw each other so infrequently that they only could maintain a ritualistic, purely symbolic relationship. They had too little contact even to establish the easygoing, friendly relationship that is the basis for the companionate style. Their relationship can best be called *remote.* At the other extreme, we found some grandparents who took on an active role in rearing some of their grandchildren, frequently behaving more like parents than grandparents. They tended to be in almost daily contact with their grandchildren, often after a disruptive event such as an out-of-wedlock birth, a death in the family, or, increasingly, a divorce in the middle generation. Grandparents with this type of *involved* relationship still could be spontaneous and playful, but they also exerted substantial authority over their grandchildren, imposing definite and sometimes demanding expectations.

It also became apparent during the course of the study that a grandparent often has different types of relationships with different grandchildren. In fact, we will describe in the next chapter how grandparents tend to balance these different types of relationships. For now, let us consider some examples, drawn from our preliminary and follow-up interviews, of these three styles of grandparenting: remote, companionate, and involved.

The Remote Relationship

There was no mystery about why grandparents with remote relationships found it difficult to become more than symbolic figures in their grandchildren's lives. Most simply lived too far away, as will be demonstrated later in this chapter when the statistical results from the national survey are discussed. But a few had remote relationships despite living near the study children. Mrs. Myers, a middle-class woman, widowed some time ago, lives in a southern city. While raising her three daughters, she urged them to be self-sufficient and independent, in case they were faced with crises such as the death of a husband or a divorce. But partly because of these values, she seems to have created an emotional distance between herself and her children and grandchildren. The last time Mrs. Myers saw any of her seven grandchildren was at Thanksgiving, three and one-half months prior to the interview. Two of her daughters (and four of the grandchildren) live out of state. The third daughter—the mother of the study child, Jessica, sixteen, and two other children—lives only about ten miles away; but Mrs. Myers reported that she saw Jessica less than once every two or three months during the previous year. Mrs. Myers feels she has a "nice relationship" with her daughters, but she said, "I don't think they tell me everything when there are real problems. . . . Sometimes I bite the end of my tongue off to keep from asking questions." As for Jessica and her other grandchildren: "I don't feel real, real close with any of them. . . . Maybe I haven't handled Jessica the way I should have. . . . I should have made more time for her in my life." When asked what it has meant to her to be a grandmother, Mrs. Myers replied in formal, unemotional terms:

> Well, I'm grateful that I've lived long enough to see the children. And I'm grateful that my children are carrying out the principles, the goals, the ideals that I wanted to put into them. And I hope that my grandchildren put it into their children. You know, to lead the good life, be educated, and to continue your education long after you get out of school.

It seems clear that Mrs. Myers is an emotionally distant figure in her grandchildren's lives, much to her current regret. Her parting words to the interviewer were, "If you think of any way to help me deal with my granddaughter, please let me know." Mrs. Myers's difficulty stems less from geographical distance (although that is a problem she faces with the families of her other two daughters) than from her formal, reserved style of interaction with her children and grandchildren. It is a style that we rarely observed—a throwback, perhaps, to the reserved style that the respondents attributed to *their* grandparents. But Mrs. Myers, who is after all a grandmother in the 1980s, is not satisfied any longer with the kind of relationships she has created.

The Companionate Relationship

Mrs. Winters, a seventy-year-old black woman who was interviewed at a senior citizen center, told an interviewer:

> Well, I'll tell you about grandparents. They do some extra loving. Especially when you don't have to just do everything and aren't busy [with children]—you know, spanking them and getting them going off to school, or whatever. You don't have that responsibility, so you have more love to spare; when you have grandchildren, you have more love to spare. Because the discipline goes to the parents and whoever's in charge. But you just have extra love and you will tend to spoil them a little bit. And you know, you give.

Although she emphasized love and companionship, Mrs. Winters also recognized that part of the meaning of grandparenthood is bound up in the continuity of the generations:

> Every child should have a grandparent, and I think that's the best part of your life. You know, it's just, I guess, reproduction or whatever it is, that you see your life going on like that. And increasing the family tree. And so it's something beautiful to look at.

Still, without much responsibility for the grandchildren, grandparents such as Mrs. Winters are free to focus on pleasurable relations. There was an overwhelming, unbounded character to the "extra love" she and other grandparents described, almost like romantic love. Mrs. Winters continued:

> I wonder why grandparents love their grandchildren like

> that. When you get to be a grandparent you'll understand, but I don't think it can be told to you in words. You'll understand; you don't know where that love comes from.

Mrs. Winters, like many others we interviewed, seemed to welcome the opportunity to leave the rearing of her grandchildren to their parents. As was noted in chapter 2, some observers question whether this is an honest response, whether upon introspection grandparents would admit that their lack of authority is deeply unsatisfying. Perhaps some might; but most of those we interviewed appeared to approve enthusiastically of the common arrangement. Here is Mrs. Winters again:

> The best part of being a grandmother is just that you can say, "Come here," and then, "Go over there, go, go." And they go. I mean, you can love them and then say, "Here, take them now, go on home." You know, something like that. See, the responsibility, all that responsibility, is not there. So you can take them whenever you feel that, you babysit when you feel like it, and then you can go. It's nice.

This is said by someone who obviously cares a great deal about her grandchildren and even lives with two of them. Yet responsibility is not something she wants.

The "I can love them and send them home" motif was repeated often in the face-to-face interviews. In our best judgment, these responses were honest and deeply felt. Having raised their own children, grandparents are ready to leave the tough work of parenting to the parents. They have paid their dues, and they are now approaching (or in) retirement. They feel that this is the time to enjoy life, to pass on to others the duties of working and parenting. In the past, far fewer people were fortunate enough to experience a financially comfortable, extended period of old age. Fewer had the luxury of leading an independent life. Now that this experience is widespread, grandparents are choosing to make leisure rather than labor the basis of their relationships with their grandchildren.

Nevertheless, grandparents also know that even if they wanted more authority, it would be difficult for them to exercise it. For they subscribe to the widely held belief that grandparents ought not to interfere in the ways their children are raising the grandchildren. This "norm of noninterference," as we will call it, is recognized as a central feature of the relationship between older parents and their adult children in the United States.[1] Grandparents are loath to violate this powerful norm; "interfering" is seen as one of the worst sins a grandparent can commit. The power of the norm reflects the ascendency of the husband-wife bond over the parent-child bond; parents have no "right" to tell their married children what to do. As one grandmother explained:

> Quite often it is said that the mother and father would still be together if the mother-in-law had not interfered. So I feel that situations between a mother and a father are very, very touchy situations. I feel that the mother-in-law has to be very careful, very tactful, how she gets in the picture, if she gets in the picture at all. I feel that unless it's . . . a problem developing into a situation where the daughter or the granddaughter's life may be at stake, or something detrimental might be done to them, I think if possible the grandparent should stay out of it.

Grandparents with companionate relationships may want to speak out, but they do not feel that they have the authority to act like parents. They must learn the limits of proper grandparental behavior. The response of one grandmother in a follow-up interview was typical:

INTERVIEWER: *How often have you felt like, let's say, expressing a difference of opinion, but didn't, about the way [the study child's] parents were bringing her up?*

Well, everybody has their own opinion as to how they raise children. . . . They'd be strict with her if she didn't clean her plate . . . and I personally didn't think, I was never one to think, well, they have to finish what's on their plate.

INTERVIEWER: *But then there are moments when you sort of hold your tongue?*

Oh, sure, you have to hold your tongue [laughter]. That's a big, important step.

It is a big, important step because it is part of the process of learning the rules of grandparenting as practiced in most families today. Most of the grandparents with whom we talked have little to say concerning the major decisions about their grandchildren's lives. Here again, grandparents rationalize their lack of influence with reference to the norm of noninterference:

They just tell me they're sending him off to camp, to some university—I don't know where it is he went last year. And I say, "Oh, that's lovely, dear." They just go along with their own life, you know. . . . I would never step into [her daughter or son-in-law's] lives. I never have done that, and [her deceased husband] didn't believe in that. Because that's how you can break up a marriage, you know?

Or consider Mr. and Mrs. Schmidt, who live in a small town in the Midwest; they have four children, ten grandchildren, and two great-grandchildren, most of whom live nearby. Years ago, Mrs. Schmidt's daughter Janice was seriously ill, and Mrs. Schmidt kept Janice's son Norman, the study child, for months at a time. Now she sees Norman, fifteen, once or twice a month. The last time the Schmidts had seen any of their grandchildren was three days prior to the interview. Despite the proximity of their extended family and the regular contact, the Schmidts are careful to keep their distance from their children's and grandchildren's lives. Mr. Schmidt explained:

> Sure we appreciate our grandchildren, and we do anything we can to help them along, and things like that. But . . . some people I think go overboard—maybe they do too much for their grandchildren, and things like this. I think they ought to be a little bit left on their own. . . . I don't think the grandparents should interfere with the parents.

There is even some suggestion that the Schmidts might feel a bit burdened by all their grandchildren and the responsibilities they entail:

> INTERVIEWER: *So, are there any other thoughts you have about being a grandmother that I didn't ask you?*

Mrs. Schmidt responded:

> I don't know what. . . . I'm just getting too many to keep track of. They're nice to have and so on, but I said when Christmas time comes, what in the world are you supposed to do?

Mrs. Schmidt's remark, even though said partly in jest, suggests some ambivalence about her relationships with her grandchildren. But for many grandparents, probably including the Schmidts, the companionate style is seen as desirable and rewarding. For example, Mrs. Waters, who lives in an eastern city, resides a block away from her daughter and two children, one of whom is the fourteen-year-old study child, Linda. Years earlier, Linda and her mother had lived with Mrs. Waters for nine months after the mother's divorce. Mrs. Waters says she is "extremely close" to Linda, whom she now sees three or four times a week. And yet her visits with Linda are brief, often momentary, as when Mrs. Waters stops by her daughter's house and Linda is going in or out. Linda, who Mrs. Waters says is "very, very busy," never calls her grandmother, nor do they sit down often and talk. Mrs. Waters explains this as normal behavior for a teenager. She does not expect more, and she is satisfied with her relationships with Linda and her older sister Rachel, seventeen. Although Mrs. Waters speaks wistfully of the time when the grandchildren were younger and would stay overnight at her home or need help with homework, she accepts the fact that those days are over because the grandchildren are older: "Now that's gone because here's Rachel, she's seventeen years old. Who wants to go and stay with their grandmother at seventeen years old [laughter]?" Still, Mrs. Waters derives great satisfaction from her past and present involvement with them. Being a grandmother, she says, has been "a terrific thing. These children have been my life." (Mrs. Waters's acceptance of the limited nature of her current relationships with adolescent grandchildren suggests that there is a life course of grandparenting, a point to which we will return in the next chapter.)

Satisfied and loving, but passive and accepting of the limitations of their relationship—these characteristics seem to describe the grandparents with companionate relationships. But just how satisfied were they? In order to probe further, the follow-up interviews included the following story:

> INTERVIEWER: *Here's a situation that happens to grandparents, and I'd just like to read it to you and get your reaction: Mrs. Smith lives a half hour's drive from her son, daughter-in-law, and two grandchildren. Mrs. Smith is unhappy because she doesn't get to see the grandchildren as much as she would like. Sometimes a few weeks go by between visits. She realizes that both her son and her daughter-in-law work full time and that the grandchildren are busy with school activities. But she thinks they*

> *could make more of an effort to see her. What, if anything, should she do about it?*

Many of the grandparents in companionate relationships were reluctant to propose action. Mrs. Schmidt laughed at the end of the story, then said:

> We've run into that right now. All the kids are into everything, you know? Like I was always used to being by myself so much, that if they can come it's all right, and if they can't, they have to live their lives.

Mrs. Schmidt takes whatever interaction she can get and does not press for more. The same story led to a lengthier discussion with Mrs. Hatfield, who lives in a small town in the Midwest:

> Just do like I do, just sit, and when they're ready, they'll come. I don't believe in begging anybody. No, I don't see my grandchildren all the time, but I know they're busy, especially at this age, teenagers. Always busy children. But I know they're there and they know I'm here.
>
> INTERVIEWER: *Do you go for two weeks without seeing them?*
>
> Oh, yes. . . . They might be busy, they have their own life too, but that doesn't mean there isn't love coming through.
>
> INTERVIEWER: *How long a time goes by when no one has called, not any of. . . .*
>
> Maybe a couple of weeks, but that doesn't bother me. Because I'm busy and they're busy.
>
> INTERVIEWER: *And I'll bet you have many friends as well.*
>
> Yeah, we have friends. And I never get my feelings hurt if I don't see them, I don't even think about anything like that, because I know that eventually we'll get together. . . . But I know they're busy. Like one of the twins, she's bowling all the time, and the other one, she bowls too. And they swim,

especially during the summer. And I don't expect them to call me. But if there's something going on and I need to know, they're there and they know I'm here.

INTERVIEWER: *It seems to work very well, you seem to have the ticket.*

I don't know, but we sure get along great. We have a good time together.

Mrs. Hatfield leaves the initiative to her children and grandchildren primarily, it seems, because she cannot do otherwise. Her grandchildren are busy, they have their own lives, and they have many other things to do. Unstated, but just beneath the surface, is the realization that Mrs. Hatfield has no way of making the grandchildren visit her more often. Her authority to make such a demand is limited by prevailing norms about the independence of children and grandchildren and by the plain fact that there is nothing crucial she can withhold or give to make them do as she would wish. As a companionate grandmother, she is powerless to demand more access to them. Consequently, to ask for more frequent contact would be demeaning ("I don't believe in begging") and in all likelihood, unsuccessful. Such a request would only strain her relationships. In order to maximize her emotional attachment to her grandchildren—which seems to be her goal—Mrs. Hatfield therefore accepts the current situation as the best that can be expected, rather than pressing for more.

Despite their limitations, Mrs. Hatfield seems to feel genuinely secure about her relationships with her grandchildren. There is still love coming through, she says. The grandchildren eventually will come around to visit. They know she is there if they need her. Moreover, they are teenagers, and teenagers are more involved with their peers. Mrs. Hatfield seems to be reconciled, at the least, to being both connected to and separate from her grandchildren. She feels emotionally close to them and accepts as natural and inevitable the restrictions on the amount of interaction they have.

Are Mrs. Hatfield and the Schmidts and other companionate grandparents satisfied with their relationships with their grandchildren? If you ask them directly, they say yes. Even under closer questioning, they maintain that the current state of affairs is satisfying ("We sure get along great; we have a good time together") and even preferable to greater involvement ("Some people I think go overboard, maybe they do too much for their grandchildren"). It is also clear, however, that some—such as Mrs. Hatfield—would like more interaction but know they are powerless to get it. Grandparents such as Mrs. Hatfield have a sliding scale of expectations that is adjusted up or down to fit the reality of what they can obtain. ("I don't see my grandchildren all the time, but I know they're busy.") Yet most companionate grandparents insist that their relationships are emotionally rewarding. Critics may think they are fooling themselves, that these are not meaningful, vital relationships, but these grandparents by and large believe that they are, indeed, satisfied. In part, they are satisfied because they compare themselves to other grandparents who they believe get even less satisfaction—as when one of the grandparents at the Jewish senior citizen center contrasted herself with the "hidden ones" who won't say anything because there's nothing good to say.

In addition, it is possible that some of the companionate grandparents have more influence on their adult children's and grandchildren's lives than they admit. That they acknowledge the norm of noninterference is clear. But in order to learn whether they always adhere to it in practice, we would need more intensive, observational studies of the interactions among the three generations. Had we talked further with the middle generation, we might have found that grandparents exert influence in subtle ways. In the next chapter, for example, we will note how some grandparents use gifts to help maintain problematic relationships. We would not be surprised if larger transfers of money, such as a partial downpayment on a house, are linked in the minds of parents and children to the maintenance of relationships that suit the older generation's values and needs. The childcare services that many grandparents perform when the grandchildren are young also may yield influence, as may being available merely to consult or to offer helpful suggestions. Without close study of family interactions, we cannot exclude the possibility that although parents have all the formal power, companionate grandparents exert more informal influence than they say.

The Involved Relationship

Mr. Sampson, who at age seventy-six has been a widower for thirteen years, lives in a lower-middle-class neighborhood of a midwestern city. Our follow-up interview with him was conducted at a cluttered dining room table filled with books, papers, correspondence, tools, and a typewriter. He was quick to explain that having been retired for a long time, he has pursued lots of hobbies and interests and that he belongs to several fraternal organizations. His grandchild Bob, the study child, has recently gone off to college. Bob and his family live just two blocks away. Before Bob left for college, Mr. Sampson typically saw him two or three times a week.

INTERVIEWER: *So what happens when Bob comes over?*

Well, we just talk about anything that comes up, or he comes over to get some help, or I do something for him, I want him to help me or something. You know, we used to take a tremendous number of trips together and things. I've had him up into Canada, I've had him down in Florida, I've had him out at the lake.

INTERVIEWER: *How do you usually spend your time? Does he just come and sort of hang around and talk?*

Mostly just talk. Unless there's something we want to fix up and all. I used to fix a radio and some of these things for him, same way that his dad and I work together on some of these things. And if he has a problem, he'll come over to see me. . . . And if I need help, like getting some screens down or putting screens in for the summer, taking the windows out for the winter, I'll get him to help me bring these down. So the connection there, it just depends on what turns up.

INTERVIEWER: *How close would you say you are with Bob?*

Real close, real close. In fact, all three grandchildren and I have had a very close relationship.

INTERVIEWER: *What do you think that comes from?*

I don't know. It's just the fact that we kind of grew up together, and grew up working with each other, and playing with each other. And one thing, of course, might have made some difference is I retired some time back—about two years after I retired my wife died, and I've been by myself, which gave me more time to spend with the boys. So when they were growing up I did spend a tremendous amount of time with them.

Lots of time together, lots of activities, lots of helping each other—these characteristics define Mr. Sampson's relationship with Bob. He also does not hesitate to discuss Bob's problems and to give Bob advice, "like about his relationship in college, that he's going to watch [out for] smoking and dope and stuff like that." And if any of his grandchildren do something he disapproves of, "I will very nicely tell them I don't think it's right. And they have never resented that too much or anything."

Mrs. Rice, another grandparent with an involved relationship, lives in a predominantly black suburb of a large western city. Since her daughter's divorce, she has helped care for her grandson Darrel, fifteen, the study child. Now she sees Darrel only about once or twice a month, but when Darrel's mother is busy for an evening or a weekend, Mrs. Rice will keep Darrel at her house and then drive him to school. Mrs. Rice definitely is not one to bite her tongue:

> INTERVIEWER: *How about when you have a difference of opinion with [your daughter] about something, do you usually tell her or do you sort of bite your tongue?*
>
> No, I'll tell her.
>
> INTERVIEWER: *How about when you see Darrel do something you disapprove of, do you . . .*
>
> I'll tell him.

She is also consulted by Darrel's mother about major decisions:

> For instance, when she took him—when he came out of public school and went into private school, thinks like that. She talked about it because she had to pay the tuition, things like that. And she thought it was better for him. And . . . she'd talk about college things . . . educational things, she'd talk about, and church activities, religious activities, and different things. Even in his sports activities, she's talked it over with me.

Some of the grandparents in involved relationships were living with their grandchildren, often after their daughters' marriages had broken up. These grandparents often took on the role of a surrogate parent—an intense, rewarding experience, to judge from the interviews, but one that also could be burdensome. When asked what it was like to live with her grandchildren, one grandmother replied, "Well, it's heaven and a hassle,

I guess you'd put it." Mrs. Williamson, a sixty-three-year-old black grandmother from a midwestern city who lived with her daughter and her granddaughter Susan, sixteen, described a typical day:

> In the morning, Susan's mom is the first to leave the house. . . . Sometimes she will wake Susan up before leaving. If not, she will say, "Mom, don't forget to wake Susan up!" So I will make sure that Susan is up. I prepare Susan's breakfast. Mornings that I have to be at work by eight o'clock, I will leave Susan here; she knows what time she is to catch her bus. . . . When we come in in the evening . . . Susan actually gets in about ten minutes till four. Her mom and I get in about four thirty. . . . I prepare all of the meals. . . . I am the one that will insist that Susan eat a good meal, take her vitamin. Susan will do the dishes. Then, after that, there's a period of looking at television, then Susan will get into her books. And she is going to finish that homework before going to bed.

Susan, Mrs. Williamson says, "will ask my opinion quite often before asking her mom's." Being a grandparent is "wonderful," Mrs. Williamson told the interviewer, and there were no disadvantages she could think of to having her granddaughter live with her.

When grandparents in involved relationships were read the story about Mrs. Smith, the grandmother who is unhappy because she does not get to see enough of her grandchildren, they tended to recommend direct action. Unlike the grandparents in companionate relationships, they did not see why Mrs. Smith should just sit tight and wait until her grandchildren came around. Mrs. Rice replied:

> Hmmm, let's see, what could she do about that. She could

> give them a call, and if they're home, run over to see them [laughter].
>
> INTERVIEWER: *So she should make more of an effort?*
>
> I think so. They are busy, maybe she could talk to the parents and tell them to come on over, we're going to have a cookout. You know, she could invite them over. Find out what day—nobody's busy every day—find out what days they are not busy and that particular day put forth an effort to invite them over or go over to their place.

The solution seems simple to Mrs. Rice; there is no thought that this course of action might be seen as interfering. Mr. Sampson noted that even if Mrs. Smith does not want to interfere in her children's lives, she could still communicate directly with the grandchildren:

> I think sometimes in a case of that kind there could be a little more effort done on that. But possibly the grandparents could help that by seeing the children without interfering with [the parents], taking their time. In other words, the grandparents could possibly go and see them. As long as there wasn't any objection to doing that; I wouldn't think from what you said there would be an objection. It's just that the parents were tied up and didn't have time to do it. But maybe the grandparents would have time to do something.

Mr. Sampson recommends that the unhappy grandparent circumvent the parents, but this is precisely what most companionate grandparents are reluctant to do. Moreover, he dismisses the possibility that the grandchildren might be too busy with school or other activities to want to see more of their grandparents; rather, "it's just that the parents were tied up." Grandparents like Mrs. Rice and Mr. Sampson matter-of-factly recommend direct action because that is the way they manage their own involved relationships with their grandchildren. The authority they develop to advise and discipline their grandchildren and their involvement in the grandchildren's day-to-day lives gives them license to ignore the norm of noninterference.

But even involved grandparents must modify their relationships as the grandchildren grow up. Mr. Sampson was well aware that Bob was becoming an adult, and he seemed to have come to terms with the impending change this would bring to their relationship. When asked what it meant to him to be a grandfather, he replied:

> I would hate like thunder not to be one. . . . No, to me it is really part of my life, and I would miss a terrible lot of activity, a terrible lot of pleasure and everything else, if I did not have grandchildren. . . . All the trips we used to take—it was fun for me, I enjoyed it. . . . I wanted to do it. . . . And I would miss that if it weren't. Of course, I'm missing it now, but I realize that I can't keep up this activity because they have their own lives to lead. And I think that's one thing, grandparents sometimes make a slip on that: they don't realize that the kids are growing up. You've got your own life to live, I don't care what it is. I've lived mine, they've got theirs coming up. Of course I may miss it; I do. I miss the boy because he's away, I don't see him. On the other hand, to help that I get into a tremendous amount of activities of my own.

The relationship is changing, Mr. Sampson implies, not because of any emotional distance, but rather that for older adolescents, growing away is part of growing up. Just as parents must learn to accept the increasing independence of their children as they approach adulthood, so too must grandparents—even grandparents with relationships as strong and deep as Mr. Sampson's.

Grandparenting: A General View

Remote, companionate, involved: these styles of grandparenting seemed to capture the major variations in grandparent-grandchild relationships that were described to us in face-to-face interviews. But were the grandparents whom we managed to interview in person representative of the national population? Can we generalize these observations to grandparents in the country as a whole? To answer these questions we turn to the data from the national telephone survey of grandparents. In order to test the generalizability of our observations, we classified the different styles of relationships between the grandparents and the study children in the national survey. There have been several attempts to classify the styles and meanings of being a grandparent.[2] All find a diversity of responses that fall in a continuum from remoteness to substantial involvement. The previous studies have provided much useful information, but they also have tended to be small, exploratory studies that were geographically, socially, and ethnically limited. They therefore cannot tell us whether styles of grandparenting vary systematically by age, ethnicity, or other social and economic characteristics. Moreover, these studies leave us with a rather static view of grandparenting, as if we could pin a label on a grandmother shortly after her first grandchild was born ("fun-seeker" or "distant figure") and be sure that the label would remain accurate for all her grandchildren for the rest of her life. In the remainder of this chapter, we will examine what proportion of grandparent-grandchild relationships in the national survey fit the remote, companionate, and involved styles. In subsequent chapters we will look at issues that go beyond simple classification, such as how grandparent-grandchild relationships change during the grandparental career, how grandparents maintain different kinds of relationships with different grandchildren, and how the social and economic characteristics of families affect these relationships.

Classification Procedures

Two Caveats. First, as noted previously, this classification refers to a particular relationship, not to a person. A grandparent who has a remote relationship with the study child might well have an involved relationship with another grandchild not in our study. Second, any classificatory exercise such as this is somewhat arbitrary. The reader should not take the categories presented in the following table and the percentages attached to them as cast in stone. They unavoidably reflect subjective, though defensible, decisions about when the nature of a relationship changes. The point is not to create a rigid set of categories that are valid for all time but rather to uncover the major varieties of grandparent-grandchild relationships and to estimate the approximate proportion of relationships that fall into each category.

Degree of Contact

Our classification procedure was based first of all on a principle that emerged from our face-to-face interviews: most grandparents and grandchildren cannot establish anything more than a ritualistic, symbolic bond unless they see each other with some regularity. As it turns out, most grandparents and grandchildren do see each other with some regularity. When we asked the grandparents in the survey how often they had seen the study child in the past twelve months, these were the responses:

Almost every day	12
Two or three times a week	11
About once a week	15
Once or twice a month	20
Once every two or three months	13
Less often	20
Not at all	9
	100%

Nearly four out of ten saw the study child at least once per week. An additional two out of ten saw the study child once or twice a month. Most grandparents, then, were not isolated from the grandchildren in the study. Recall, however, that the names of the grandparents in the survey were provided by the parents, who undoubtedly were less willing to identify grandparents from whom they were estranged. So the amount of contact probably is somewhat higher than in the general population. Nevertheless, the overall frequency of contact is impressive, especially given that many of the grandparents—as we will show in the next chapter—had more contact with other grandchildren than with the study children.

Still, the responses also show that there is a wide variation in the amount of contact. Nearly three out of ten grandparents answered that they had seen the study child "less often" than once every two or three months. Clearly, there is no single pattern of visitation that holds for all grandparents and all grandchildren. One intriguing question, which we will explore in chapter 5, is why so much variation exists. For now, we will note only that distance has a great deal to do with it: in 82 percent of the cases where visits had occurred less often than once every two or three months, the grandparents lived more than one hundred miles from the study children. Thus, distance created a geographical barrier to contact and stronger ties. We classified the 29 percent of grandparents who saw the study children less often than once every two or three months as having remote relationships with the study children.

Grandparents who reported more frequent contact with the study child in the past twelve months obviously had a better chance to establish the affectionate, informal type of interaction that we have called the companionate style. Most, it appears, succeeded in doing so. The survey included a list of eleven activities (questions 26a through 26k in appendix 1), drawn from our preliminary interviews, that grandparents sometimes do with their grandchildren. The interviewers asked the grandparents if they had done each of these activities during the past twelve months. (Obviously, it was not possible to ask these questions of grandparents who had not seen the study child in the past twelve months.) The results are displayed in figure 3–1. Leisure-oriented, informal activities were cited by most grandparents: joking, kidding around, watching television, reminiscing about the old days. And four out of five reported that they gave money to the grandchild. These activities fit the pattern of a relationship based on easy informality and companionship. Smaller, but still substantial, proportions reported giving advice or discussing the grandchild's problems.

Only a minority of grandparents, however, reported going to church or synagogue with the grandchild, taking a day trip together, or teaching a skill or game. The grandparents were quick to point out that they do not do more with their teenaged grandchildren because the grandchildren are busy with school activities, work, and friends—in other words, because they are growing up and away from their families.

> You always feel like you wished you could have spent just a little more time with them. But when they have things to do, and other things that they do on their own, you can't expect them to be with you. Heck, they want to be with young people, you know.

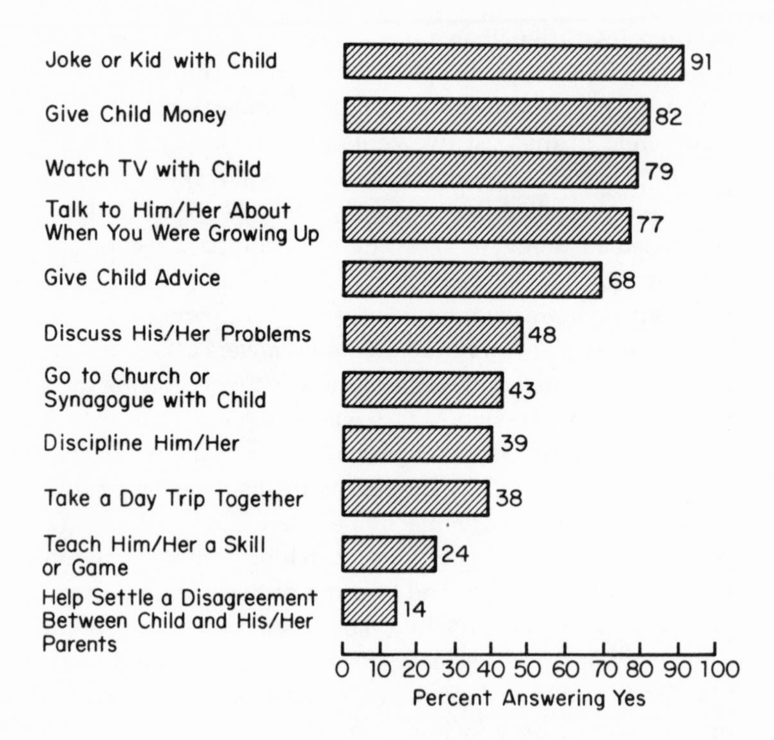

FIGURE 3–1

Percentage Engaging in Various Activities with their Grandchildren During the Previous Twelve Months

There are some things that most grandparents emphatically do not do with their grandchildren. In particular, all but a distinctive, small group avoid stepping between parents and children. Just 14 percent, as figure 3–1 shows, said that they had helped settle a disagreement in the past year between the study child and his parents.

Extent of Influence

The dominant style, then, seems to be based on pleasurable interaction and affection. But a further examination of these and other survey questions suggests that a minority of relationships encompassed something additional. We used the statistical technique of factor analysis to identify distinct clusters of activities that only some grandparents and grandchildren engaged in. Two such clusters emerged. The first revealed a pattern of grandparents and grandchildren exchanging services. The answers to four survey questions (numbers 27a, 27b, 28a, and 28b) identified grandparents and grandchildren who helped each other with errands and chores or who asked for help with something they were doing or making during the previous year. We formed an "exchange" scale with a range of zero to four by summing the number of positive responses to these four questions. Most grandparents and grandchildren had a limited exchange of services; but about one-third answered positively to at least three of the four questions, and about one-sixth answered positively to all four.

The second cluster of activities that emerged in our analysis measured the extent to which the grandparents were able to exert the type of influence over the grandchild that is typically reserved for parents. The cluster was composed of the responses to five questions (26f, 26g, 26j, 32, and 33). Given the importance of these questions to our subsequent analysis, let us quote them verbatim:

1. During the past twelve months, did you discipline [him/her]?
2. During the past twelve months, did you give [child] advice?
3. During the past twelve months, did you discuss [his/her] problems?

4. When you see [child] do something you disapprove of, do you correct [him/her] often, sometimes, hardly ever, or never?
5. Do your children consult you before making an important decision about [child] often, sometimes, hardly ever, or never?

A scale of "parentlike" behavior with a range of zero to five was formed by summing the number of positive responses to the first three items and the number of "often" or "sometimes" responses to the latter two items. The most common score was zero and the least common was five, reflecting the lack of authority in these matters among most grandparents. For example, only 39 percent reported ever disciplining the child. Only 21 percent were consulted "often" or "sometimes" about important decisions. Only 45 percent corrected the child "often" or "sometimes" when he or she did something of which they disapproved. Still, nearly one-half of the grandparents scored three or more and about one-fourth scored four or more. Grandparents with high scores on this scale are of particular interest, for they have been able to surmount, at least partially, the powerful norm of noninterference. Grandparents who are disciplining, advising, and correcting their grandchildren with regularity have crossed a sensitive boundary in intergenerational relations.

Classifying the Intergenerational Relationship

We therefore sought to separate those relationships with a strong component of exchange and parentlike behavior from those that were primarily companionate in nature. To do so we first divided the scores on each scale into two parts: scores of zero, one, two, or three on the scale of parentlike behavior were considered low; scores of four or five were considered high. Similarly, scores of zero, one, or two on the exchange scale were considered low; scores of three or four were considered high. Thus, in order to score high on either scale, a grandparent had to respond positively to most of the relevant questions; this rather stringent rule reflected our impression that some grandparents tended to exaggerate the amount of exchange or influence they experienced. We then defined a companionate relationship as one in which the grandparent and grandchild had seen each other at least once every two or three months during the past year but in which there were low levels of either exchanges of services or parentlike behavior. An involved relationship was defined as one in which the grandparent and grandchild had seen each other at least once every two or three months during the past year (most, in fact, saw each other much more often than that) and in which there were high levels of both exchanges and parentlike behavior. These distinctions yielded the following three-way classification of the relationships between the grandparents and the study children in our national sample:

Remote	29
Companionate	55
Involved	16
	100%

As mentioned previously, these percentages should not be taken too literally because other reasonable cutoff points would alter them. Moreover, they are approximations: some grandparents in remote relationships may see their grandchildren infrequently but for long, intense visits, as during school vacations; and a few companionate relationships may have been lacking in affection and spontaneity. In addition, as discussed in appendix 1, our sampling limitations probably led to a modest underestimate of the size of the remote group and a modest overestimate of the companionate and involved groups.

Still, the three-way distinction of remote, companionate, and involved relationships seemed to capture real differences in the lives of grandparents and grandchildren. In the survey, all grandparents were asked whether their relationship with the study child was extremely close, quite close, fairly close, or not very close (question 25); and whether the grandparent and the study child could "exchange ideas about things that really mattered" extremely well, quite well, fairly well, or not very well (question 24). Experience has shown that when survey respondents are asked these kinds of questions about their family lives, they reply in very positive terms.[3] Granted, most people do feel close to their kin and are satisfied with their family lives. But it is also very embarrassing to admit to a stranger that in this central aspect of life one's relationships are deficient. It may also be difficult to admit to *oneself* that family relations are unsatisfactory. So, it is noteworthy that 21 percent of grandparents in remote relationships were willing to admit that they were "not very close" to their grandchild, compared to only 4 percent of those in companionate relationships and 2 percent of those in involved relationships. Similarly, 22 percent of the remote grandparents said that they could exchange ideas or talk about things that mattered "not very well," compared to 9 percent of the companionate grandparents and 2 percent of the involved grandparents.

There were also sharp differences in how satisfied the grandparents were with the amount of time they spent with the study child (question 19). Only 10 percent of the remote grandparents said they spent "about the right amount of time," as opposed to 32 percent of the companionate grandparents and 52 percent of the involved grandparents. In fact, 75 percent of the remote grandparents responded that they spent "a lot less time than [they] would like" with the study child, compared to 28 percent of the companionate grandparents and 23 percent of the involved grandparents. The sources of this dissatisfaction became clearer when the interviewers presented a list of possible reasons with the introduction, "Do you ever find that you don't spend as much time with [the study child] as you would like because of the following?" Nearly all remote grandparents—93 percent—agreed that the study child "lives too far away," compared to 38 percent of the companionate grandparents and 19 percent of the involved grandparents. (Recall that most of the remote grandparents lived more than one hundred miles from the study child.) Sixty-four percent of the remote grandparents agreed that "the trip is too difficult." Thus, distance once again emerged as the primary barrier to more frequent contact in remote relationships—a point to which we will return in chapter 5. The major complaint of the companionate and involved grandparents, in contrast, was that the child is "too busy" (about half of the companionate and involved grandparents agreed with this statement) or that the child's parents are "too busy" (about one-third and one-fifth, respectively, agreed). The "too busy" theme emerged often in our follow-up interviews, and we will have more to say about it.

Before leaving this series of questions, it is important to add that 21 percent of the remote and companionate grandparents, and 12 percent of the involved grandparents, agreed with the statement that their health prevented more frequent visits. Thus, health problems did limit the relationships of some grandparents. Only 1 percent of all grandparents, however, agreed that either of the child's parents did not want the grandparent to see the child more often. This tiny response is noteworthy, given the presumption of some "grandparents' rights" activists that large numbers of grandparents are prevented from seeing their grandchildren by their ex-daughters-in-law following a divorce. To be sure, our selection process may have screened out some grandparents whose access to the children was barred. (The process also may have screened out some of the least healthy grandparents.) We did uncover a few such cases in our face-to-face interviews. But the responses of the national sample we did obtain cast doubt on the notion that lots of grandparents see their grandchildren infrequently because they are forbidden to do otherwise. Far more serious impediments to frequent contact are distance and, secondarily, the separateness implied by the common response that everyone is "too busy."

These general questions about satisfaction with emotional closeness, talking about things that really matter, and time spent together suggest that there are substantial differences among the three relationships. Although our classification scheme is necessarily approximate, any other reasonable scheme would retain the essential findings reported here: a minority of grandparents did not see the study child often enough to establish more than a symbolic, distant relationship. Nearly all of the rest were able to establish the affectionate, loving, informal relationship that we have labeled companionate. In most of these relationships, the parties exchanged few services and the grandparents left parenting to the parents. But in some cases, the bonds of affection were overlaid with ties based on exchange and parentlike behavior—a style we have labeled involved. Involved grandparents actually engaged in informal, leisure-oriented activities more often than those in companionate relationships—99 percent joked or kidded with the study child, 96 percent gave the child money, 94 percent watched television together, and all of them talked to the child about when they were growing up. But the involved relationships took on an instrumental character as well.

Variations

The statistical portrait drawn from our national survey confirms the inferences we drew earlier in the chapter from the personal interviews. The classification of grandparent-grandchild relationships into three styles seems to fit American grandparents fairly well. Nevertheless, this distinction, though helpful, is too static. Recall Mr. Sampson's remarks about how the relationships between grandparents and grandchildren change as both get older. Moreover, grandparents can have different kinds of relationships with different grandchildren simultaneously. These variations through time and space are not random; rather, they reflect different stages of life, the constraints of distance, the exigencies of major life events, and personal preferences. As we learned more about these variations, we began to think of grandparenting as a career—a role with distinct stages that change over time—with opportunities for investment and with constraints on activities. Grandparents, we observed, try to organize their careers so as to maximize the satisfaction they obtain from their role. In the next chapter, we examine grandparental careers and their implications for American kinship.

In addition, the information presented in this chapter has pointed to some important sources of differences in grandparent-grandchild relations, such as geographical distance and frequency of contact. And it has hinted at some others, such as the grandparent's relationship with the parents, marital disruption, and ethnicity. There are other plausible sources of variation about which this chapter has been silent: gender and social class, for example. In chapter 5 we will look more systematically at the sources of variation in the roles and relationships of American grandparents.

4

Grandparental Careers

Grandparents have careers, not unlike the occupational careers that constitute our work lives. Of course, no one is hired to be a grandparent, no one gets paid for it, and few people think of it as work. Being a grandparent doesn't necessarily involve the kind of dedication and involvement we associate with moving up the corporate ladder. Yet the similarities exist, and an examination of them can help improve our understanding of the role grandparents play in American kinship.

Like a career, grandparenthood has distinct stages. The career begins when the first grandchild is born, and it remains in its initial stage until the grandchildren reach adolescence. During this period, grandparents establish companionate relationships with their grandchildren and are called upon typically for babysitting, help with homework, and other kinds of direct assistance. As in a career, grandparents invest heavily when the grandchildren are young—in terms of both time and emotional commitment—and hope to enjoy continuing interest from this investment later on. Then, in the second stage, the teenaged grandchildren begin to break away from their families, including their grandparents. Babysitting and other services are not needed as often; moreover, the grandparents, who are aging, find it more difficult to provide frequent services. In fact, teenaged grandchildren often provide help to their grandparents with errands or chores. The third stage emerges when the grandchildren reach adulthood and start their own families. Then the grandparents must relinquish their role to their own children, becoming instead great-grandparents—a role with limited, symbolic meaning and little content. Thus, as grandparents and their grandchildren grow older, the nature of the relationship changes. The grandparental career follows a predictable course, beginning with a more active and engaged phase and concluding with a more honorific and ceremonial phase. For most grandparents, these stages overlap because there are several grandchildren of varying ages. But it is still useful to think of the stages as analytically distinct.

Preadolescence

In every home we visited, photographs of the grandchildren either were on display or were happily provided upon request. Most grandparents had several photographs of the grandchild in our study taken at different ages. But the photographs were concentrated in the preadolescent years—infancy, toddler age, elementary school—even though the grandchildren in the study were now teenagers. The numerous snapshots of the preteen years suggested that this may have been an especially favored time for the grandparents. Indeed, one grandmother referred to the preteen years as "the fat part of grandparenting."

Our survey results confirm this preference for the early years. We asked the respondents how old the study child was when they most enjoyed being his or her grandparent (question 90). Forty-two percent replied that all ages were equally enjoyable. This is, to be sure, a reasonable response, but we suspect that some grandparents were giving what they believed to be socially appropriate answers. After all, selecting a most enjoyable age implies that the grandchild was less enjoyable at other ages; and some grandparents may have felt uncomfortable making such an implicitly negative judgment to a stranger (or to themselves). Of those who did state a preference, the early years were overwhelmingly favored. Here are the results, presented according to broad age ranges:[1]

All ages equally	42
Preschool (0–4 years)	23
Elementary (5–12 years)	12
Preschool and elementary (0–12 years)	14
Teenage (13–18 years)	8
Elementary and teenage (5–18 years)	2
	101%

Nearly one-fourth selected the preschool years, which had occurred at least a decade previously, as the most enjoyable time. Significantly, very few favored the present period, at the time of the interview, when the study children were teenagers.

The early years of the grandchild's life were recalled fondly by our grandparents. They remembered the excitement of just becoming a grandparent (or becoming one again) and of witnessing the extension of the chain of being to the next generation. Most of the grandparents were heavily involved in the newly formed families, often visiting their children at the time of birth to help the young parents adjust to their new responsibilities. One-fourth of the grandparents in our survey helped to name some of their grandchildren. Others provided childcare instruction and advice. By these acts, the grandparents acknowledged the creation of the new family and the succession of the generations. They formed a significant audience for the parents, helping them to establish their new identities.

The early years also are remembered fondly because very

small children are a source of delight, particularly if you can hand them back to their parents when their diapers need changing. Young children can be held, hugged, played with, and enjoyed purely as objects of affection. Most respond enthusiastically to the doting attention of grandparents. Perhaps this is why Neugarten and Weinstein, in their classic study of styles of grandparenting, found the "fun-seeker" pattern most common among grandparents under sixty-five.[2] The fun-seeker, they wrote, "joins the child in specific activities for the specific purpose of having fun, somewhat as if he were the child's playmate." The authors report fewer fun-seekers among grandparents over sixty-five, a difference they attribute to the aging process or to trends over time in values and expectations. But it seems more likely to us that the difference primarily reflects the aging of the grandchildren. The fun-seeker pattern failed to emerge in our data not because our grandparents did not like fun but rather because it was not an appropriate style with older grandchildren: no matter how deep and warm the relationship remains over time, a grandmother does not bounce a teenager on her knee.

In addition, grandparents can provide direct assistance more easily when the grandchildren are young. Babysitting is the major example; the grandparents in our study often babysat for their younger grandchildren. One woman who was providing babysitting services at the time of the follow-up interview said:

> [My daughter] has been working for me at the shop, and I've been babysitting these kids for the last two years. It is going to be really different not seeing them every day. Sometimes they get under my skin, but it's like an abscess, you'll miss it once it's gone, or a toothache [turns to granddaughter], huh, Janice? . . . Hey, I'm a prisoner, you know, I'm the dumping ground [laughs]. You know, you always think, boy, when my children are grown I can have a decent rug

> and decent furniture. Oh no, you start all over again with grandchildren.

Despite these mock complaints, it is clear that she will miss having Janice around when the babysitting arrangement ends. When Janice is older, her grandmother's assistance will not be needed as much. The grandmother finally will be able to buy decent furniture, but at the cost of a less intensive relationship with her granddaughter. A similar story could be told about helping grandchildren with homework. When the grandchildren are younger, there is both a companionate role and a limited functional role for most grandparents.

Adolescence

The most common style among the grandparents over sixty-five in the Neugarten and Weinstein study was "formal." These grandparents had relatively frequent contact with the grandchild, but:

> Although they like to provide special treats and indulgences for the grandchild, and although they may occasionally take on a minor service such as babysitting, they maintain clearly demarcated lines between parenting and grandparenting, and they leave parenting strictly to the parent. They maintain a constant interest in the child but are careful not to offer advice on childrearing.[3]

These grandparents sound similar to the grandparents with companionate relationships in our study, though more reserved. Indeed, we would suggest that the emergence of the "formal" pattern reflects the fact that grandparents over sixty-five tend to have older grandchildren. Styles of grandparenting, in other words, can change as grandchildren and grandparents age. As was shown in chapter 3, grandparents and teenaged grandchildren commonly share leisure-time activities like watching television, reminiscing, and joking. To be sure, these are pleasurable activities, but they differ from the fun-seeking, game-playing, patty-cake-type activities one would do with a toddler.

In fact, teenaged grandchildren are old enough to reciprocate for the assistance they received earlier on from their grandparents. Fifty-nine percent of the grandparents with companionate relationships reported that the study child ran errands or did chores for them during the previous year. And 34 percent said that they had asked the child for help with something they were doing or making in the previous year. Among grandparents in involved relationships, these forms of assistance were (by definition) nearly universal: 93 percent reported help with errands or chores, and 95 percent asked for help with something they were doing or making. Moreover, the exchange of services sometimes flowed both ways. Among companionate grandparents, 28 percent reported that the child asked them for help with something the child was doing or making, and 33 percent reported helping the child with her errands or chores. The comparable figures for involved grandparents were 81 and 92 percent, respectively. Mr. Sampson, the involved grandparent introduced in chapter 3, talked about his grandsons who recently had gone off to college:

> If I wanted something and I had to have somebody help me, they were very good at that. And by the same token, for instance the older boy, out at the house if they have some plumbing or something to do, I'll go out and fix it for him or help him on it. And it helps both ways on the things, because I think it's good for us to do some of that.

Teenagers may not be as immediately pleasurable as younger

children, but they are old enough to help out. The exchange of services can provide an important bond between grandparent and grandchild, one that survives the bygone days of babysitting and playing games. Yet few observers have noted the flow of services between grandparents and grandchildren—probably because most researchers have studied grandparents who have younger grandchildren.

In fact, we would suggest that when people think about grandparents, they picture them with young grandchildren—the fat part of the relationship. But grandchildren (and grandparents) age, and as they do the basis of their relationship changes. Some grandparents give and receive services from their teenaged grandchildren. In families where a crisis such as divorce has occurred, grandparents may play an active, parent-like role. But for most, this is a period of pleasant, frequent, but sometimes fleeting contact with busy teenagers. Even Mr. Sampson realized as much; recall that he said in the preceding chapter, "I realize that I can't keep up this activity because they have their own lives to lead. And I think that's one thing, grandparents sometimes make a slip on that: they don't realize that the kids are growing up." Grandparents redefine their relationships as their grandchildren proceed through adolescence. They do not admit that they are not needed as much; rather, they say that the grandchildren are "busy"—a term that seems to imply: "that's as much of a relationship as we are able to have at this time." This process is an echo of what happens to parents when their own children reach adolescence. Grandparents with teenaged grandchildren are in a similar situation to parents, and grandparents have even fewer resources they can use to retain control.

Most grandparents, like Mr. Sampson, accept this inevitable movement toward independence. Many enjoy watching their grandchildren grow up, even though they may regret the passage of time. An interviewer asked Mrs. Hatfield, whose comments on the "Mrs. Smith" story were presented in chapter 3, whether it was true that grandparents cannot have a closer relationship with teenagers. She replied:

> I don't think so. We sit and yap, don't we? We have a great time. I like my granddaughters; we get along fine. They're getting to the age now where boys are more important than anything in the whole world, but that's natural. I like to listen about the boyfriends. We just went through a senior prom and they came over and showed me their dresses and so on, and that's fun, I enjoy it. We're just progressing along. Eventually I'll be a great-grandmom. That will be wonderful; I'll love it.

Adulthood

There is some suggestion in the sparse research literature on grandparenthood that adult grandchildren become closer, once again, to their grandparents, presumably when they marry and have their own children—the great-grandchildren.[4] This is the hope expressed by Mrs. Hatfield, who feels she is just progressing right along to great-grandparenthood. The idea that the teenage years are a cold spell in an otherwise warm climate was expressed by other grandparents. Mrs. Stevens, a women who lived near Mrs. Hatfield in a small town in the Midwest, had two twenty-four-year-old granddaughters (one of them a step-granddaughter).

> INTERVIEWER: *How do the two twenty-four-year-olds relate to you now?*
>
> Well, the one, of course, she's real busy, she is working full time—that's my real granddaughter. She's been going to school at night. Then she has a lot of activities on the weekends, so I don't see them that often. She came by yesterday

and we went for a walk. The other one, I hardly ever see her anymore. For a while she was living in [another state]. They are busy with their lives, so I just let them be.

INTERVIEWER: *Do you think grandparenting is something for when kids are young?*

Yes, I think so, and then all of a sudden when they are older or married or have a family or something, then they come back to you again. But there is a period in there when they just want to do their thing and they don't want. . . . That's all right, that doesn't bother me at all, I'm glad, I prefer it to be that way.

But her twenty-four-year-olds had not yet come back to her; they were still "real busy." Do attenuated relationships improve—and are close relationships maintained—when adult grandchildren start their own families? Our sample of grandparents was too young to provide definitive answers, but the information we did gather suggests little improvement. Forty percent of the people in our survey had great-grandchildren, typically a few at most. When we asked about the great-grandparental role, in both the survey and the face-to-face interviews, we were able to elicit few meaningful responses. Our respondents had less to say about being a great-grandparent than about any other topic we pursued. People who had eloquently described what it is like to be a grandparent became tongue-tied when asked about being a great-grandparent. Consider Mrs. Winters, who had spoken with so much emotion about the "extra love" she felt for her grandchildren:

INTERVIEWER: *How does it feel to be a great-grandmother?*

Oh, I don't know. Mmmm [laughter]. Just one more in the basket, one more egg in the basket. You get so many of them mixed up. I've got so many children and grandchildren, and I love them all.

Grandparental Careers

All great-grandparents in the survey were asked, "What difference, if any, is there between being a grandparent and a great-grandparent?" The question (number 71) was asked in what survey researchers call an "open-ended" way: no response categories were suggested; instead, the interviewers wrote down whatever the respondents said. This technique encourages a broad range of answers; yet 64 percent said only that there was no difference. Another 25 percent said that they saw the great-grandchildren less often than the grandchildren. There were a handful of other responses: 4 percent said the great-grandchildren made them feel older; another 4 percent said the birth of great-grandchildren was somehow more of a special event or more enjoyable than the births of their grandchildren; and 2 percent said they felt closer to their grandchildren.

Why this reticence from people who were pleased to discourse at length about their grandchildren? The first clue comes from the 25 percent who said they see less of their great-grandchildren. As a woman from a southern city explained in a follow-up interview:

> The biggest difference in being a great-grandparent is [that] usually about the time you get down to that third and fourth generation, the families are so scattered, you don't get to see them as much.

Given the great amount of geographical mobility in the United States, it is likely that some or all of the grandchildren will have moved away by the time they are adults. Moreover, most of the great-grandchildren were very young. When combined with the problem of geographical distance, this implies that many great-grandparents had been unable, at least so far, to establish much of a relationship with their great-grandchildren. A great-grandmother who lives on the outskirts of an eastern city gave the stock "no difference" answer in a follow-up interview, then with further probing told more:

INTERVIEWER: *Is there any difference between being a great-grandparent and being a grandparent?*

No. . . . I love them all.

INTERVIEWER: *How often do you get to see your great-grandchildren?*

Well, I haven't, not as often. They come down to [name of town], my son and daughter live around there. So it's been about a month and a half, two months, since I've seen them.

INTERVIEWER: *So you don't think there's any difference. . . .*

Well, yes, there's some.

INTERVIEWER: *Like what?*

Well, I'm closer to the grandchildren than I am to the great-grandchildren. . . . Most of them are just babies.

Perhaps the relationship will become more salient as the great-grandchildren grow a little older. But for many great-grandparents, two problems may interfere: first, the aforementioned problem of geographical distance; and second, the problem of deteriorating health. For by the time the great-grandchildren are older, the great-grandparent may be very old. Some observers have speculated that this extreme age gap may make it difficult for great-grandparents to get along with young, highly active great-grandchildren.[5] For example, in one follow-up interview, an informant from a small town in the East discussed her mother:

Ninety-four and she can get around, but she uses a cane. Mind? Her mind's better than mine. If I go and visit her or something and I make a mistake, she'll correct me.

INTERVIEWER: *That's great. . . . What kind of relationship does she have with your children and your grandchildren? Do they ever stop to see her or visit her?*

Yeah. Every now and then. But not too often, because she can't stand kids anymore.

INTERVIEWER: *She can't?*

No. She's at the point where when they [caretakers] come in and clean, she wants it to stay that way. Don't let the kids dirty—she doesn't like anything thrown around.

In addition, there does not seem to be much of an institutionalized role for great-grandparents. The four- (or five-) generation family is a new phenomenon, a creation of the recent revolution in adult mortality. Consequently, there is a lack of norms, of widely held beliefs, about how great-grandparents should behave. To be sure, the nature of grandparenthood also has changed in this century (as we have argued in earlier chapters), but the grandparent-parent-grandchild triad is deeply rooted in our conception of kinship. Not so for the great-grandparent–grandparent–parent–grandchild tetrad; it has not been common for long enough to find its place in our family system. As a result, when great-grandchildren are born, members of the oldest generation find that their role as grandparents has been usurped by their own children—the new grandparents—and that there is no generally accepted replacement. Great-grandparents, then, must make way for the new grandparents. Mrs. Winters sensed as much:

It's kind of hard to keep from butting in; see, I'm a great-grandparent, too, you know. And see, the grandparent, like my daughter, she's got a boy [a grandson], so I want to butt in every once in a while to tell her I believe that's her job. The great-grandparent's got to hold it sometime and let the grandparent do something. It's hard to keep from being bossy. That's my problem.

Her problem is how to step aside gracefully from the grandmother role, now occupied by her daughter. Just as parents must relinquish control when their adult children have offspring, so must grandparents make way. This process of generational succession leaves little room at the top of the pyramid for great-grandparents.

But what happens to the relationships between the new great-grandparents and the new parents—their adult grandchildren? In a well-known study of 300 three-generation families in the Minneapolis–St. Paul area, Reuben Hill and his colleagues found "extensive intergenerational visiting three generations in depth" and "high participation in common social activities" among grandparents, parents, and married, adult grandchildren. They also found frequent exchanges of services among the generations.[6] On further examination, however, the reports about grandparent-grandchild relationships in this study seem ambiguous. To be sure, in about one-third of the families, the adult grandchildren saw their grandparents at least once a week—a high level of contact. Yet nearly half of the adult grandchildren saw their grandparents only one to four times per year—even though all three generations had to reside in the Minneapolis–St. Paul area in order to qualify for the study.[7] Moreover, the grandparents provided far fewer services than they received, and most of what they did receive came from their children, not their adult grandchildren. In fact, when the grandparents needed help in a crisis, they "tended to turn first to their children (the parent generation), second to their peers, third to health and welfare agencies, fourth to private specialists, and fifth to their married grandchildren." When the adult grandchild generation needed help, they turned to their parents first, private specialists second, peers third, siblings fourth, and grandparents fifth.[8]

Consequently, even in a study with a restricted geographical scope that maximized the opportunities for visiting and the exchange of services, the pattern of visiting was highly variable and the amount of direct exchange between grandparents and adult grandchildren was small. In a more representative sample of three-generation lineages—one that allowed for geographical mobility—the average levels of visiting and exchange would be considerably lower.

This is not to say that grandparents necessarily have distant relationships with their married adult grandchildren. Some of the grandparents who are fortunate enough to reside near their grandchildren continue to have frequent visits and to share activities, as the Hill study noted. It may be true that some adult grandchildren renew and strengthen their relationships with their grandparents. Our survey cannot help us determine how many do so; to obtain that type of data, we would have to reinterview our families in a few years. But based on the relevant information we did collect and on past studies, the situation appears to be as follows: when grandchildren reach adulthood, there is little evidence of direct exchange of services and there is great variability in the frequency of visiting. Older grandparents may continue to see some nearby adult grandchildren often; but other adult grandchildren, especially those who have moved away, are seen much less often. Grandparents' relationships with their adult grandchildren, we suspect, are largely symbolic: with some exceptions among those who live nearby, grandparents see their adult grandchildren mainly on holidays or special family occasions, where the grandparents are important symbols of family continuity. Furthermore, when great-grandchildren are born, the members of the oldest generation find themselves in a truly "roleless" role: great-grandparents. Edged out by their own children, they no longer are able to play the role of grandparent with small children. These new, four-generation families deserve more detailed study. But it seems to us that for most grandparents, the entrance of grandchildren into adulthood—and particularly the birth of great-grandchildren—signals the end of the grandparental career in all but a symbolic sense.

Emotional Investment

It also became apparent during our study that, as in a career, grandparents work to construct the most satisfying role from the opportunities available to them, subject to the constraints they face. If they have several grandchildren (as did most in our sample), their relationships will differ in emotional closeness and frequency of contact. Some grandchildren may live nearby; one or two may be the child of a son or daughter with whom the grandparent is especially close; another may need assistance from grandparents after a parental divorce. Grandparents will tend to focus their attention on grandchildren such as these, and they will have less to do with other grandchildren with whom the opportunities for involvement are limited. They tend to adopt a strategy, in other words, that we will call *selective investment.* They concentrate on those relationships that promise the greatest emotional return on their investment of time and effort. This allows them to maximize the chance of establishing close, satisfying relationships with at least some of their grandchildren. Having done so, they can then generalize their feelings about these special relationships to their overall feelings about being a grandparent.

In order for this strategy to be successful, grandparents must have some grandchildren whom they see regularly. Most grandparents do. We asked all grandparents, "When was the last time you saw any grandchild?" (question 64). Grandparents who lived with the study child were automatically classified as having seen a grandchild "today." About half of the grandparents in the survey had seen a grandchild that day or the day before, and 70 percent had seen a grandchild in the past week —proportions that are consistent with results from other national surveys.[9] Just one out of six had not seen a grandchild within the previous month. Even among the grandparents in our survey who had remote relationships with the study children—who, by the definition of this category, had seen the study child less than once every two or three months during the previous year—one out of three had last seen a grandchild that day or the previous day. Thus, many of the grandparents who saw the study child infrequently had frequent contact with other grandchildren. (A tiny minority, however, were truly isolated: 6 percent of all grandparents—including one out of five remote grandparents—had last seen a grandchild more than three months previously.)

Consequently, most grandparents who had an unsatisfactory relationship with the study child could compensate by developing their relationships with other grandchildren. Sometimes the compensating relationship even could occur in the same family. Mrs. Grant, who lives in a midwestern city, does not get along well with her daughter-in-law, the mother of two of her grandchildren: William, seventeen, and the study child, Delia, thirteen. Although she lives only a short drive away, Mrs. Grant rarely sees Delia, whom she describes as close to her mother. She told the interviewer that her relationship with Delia was "not very close." Yet Mrs. Grant sees Delia's older brother, William, much more often. She feels protective toward him, for he "has always been very, well he was what you'd call one of those nervous-type babies. I went up and stayed with [the mother] after he was born." Even now, she says, William "gets on his mother's nerves." So Mrs. Grant took a particular interest in him; and now, she says, "he just loves to come here and be with his grandparents." In fact, William wants to attend the local university and live with his grandparents. Although Mrs. Grant may not feel close to Delia or her mother, she found that William needed her help and became much more actively involved in his life.

The selectivity of grandparents' involvement with their grandchildren also became apparent when we asked the grandparents the following question: "It's not unusual for grandparents to like some grandchildren more than others. Do you have a favorite grandchild?" (question 65). Because of the prevailing norm that grandparents should love all their grandchildren equally, it was difficult for the respondents to admit to a favorite. Mrs. Rice, the involved grandmother quoted in the previous chapter, told an interviewer over the telephone that a sixteen-year-old granddaughter—not the study child—was her favorite. But at the beginning of her follow-up interview three months later, she was anxious to recant:

> INTERVIEWER: *To begin, I'd like to ask you, when you finished the phone interview a couple of months ago, did you have any reactions? Anything that it made you think about?*
>
> Yes, definitely. Because after the interview I—they were asking if I had a favorite grandchild, and then I thought about it, I really don't. I try to love them all equally. But then there are some that have a greater need than others.
>
> INTERVIEWER: *But you did answer that you had a favorite grandchild?*
>
> Right, right.
>
> INTERVIEWER: *What do you think happened there, did it catch you off guard?*
>
> I think so.
>
> INTERVIEWER: *Why do you think you selected her when you were asked that?*
>
> Because I just kept her a lot when she was younger. I kept her, she lived with me off and then on. See, I'd keep her during the day when they lived in the city. And she was around me more than the others. Yes, and she is the only grandchild that I've really ever kept for . . .
>
> INTERVIEWER: *A long time?*
>
> Yes.

Still, 29 percent were willing to admit that they had a favorite. (In Mrs. Grant's case, it was William.) Common reasons given for why a particular grandchild was the favorite were that he/she had spent more time with the grandparent, had been taken care of by or lived with the grandparent, was the oldest or the youngest, or had idiosyncratic positive traits the others lacked. Fourteen percent of grandparents with involved relationships with the study child said that the study child was their favorite. (Another 21 percent stated, like Mrs. Rice, that a different grandchild was their favorite.) But among grandparents with companionate or remote relationships, few (5 percent and 4 percent, respectively) named the study child as their favorite. Instead, about one out of four companionate and remote grandparents said that they had another grandchild who was their favorite. These favorites provided compensation for other, less satisfying relationships.

Take Mrs. Allison, for instance, who lives in a suburb of a large northeastern city. She sees the study child, Christopher, once or twice a month; but when asked whether she felt close to him, she replied, "No, not really, I mean we see him a lot, but not as close as maybe some people would be to their grandson." "I really can't say we do things for him," she told an interviewer; nor could she cite much that Christopher had done for her. Mrs. Allison, however, does have a favorite: Nora, aged twenty-four, the oldest of her ten grandchildren. Nora is the child of Mrs. Allison's only daughter (she also has two sons). As to why Nora is her favorite, she said:

> I think it's because I always wanted another daughter. . . . [Nora's mother] and I were always very close, too, as she was growing up. And [Nora's mother] also has that special relationship with her daughter. Which I think is nice, that closeness there. She's always been aware of it, that Nora and I are so close. . . . From the time [Nora] was little, she practically lived here. In fact, her dad used to think that she spent too much time here.

So in this family, three generations of women feel a special bond for each other. Even though Nora is now in business school, they still keep in touch often, according to Mrs. Allison:

> She called me last Thursday, and I'll probably hear from her either tonight or tomorrow; and if I don't, I usually talk to her once a week at least. . . . With working and school—she doesn't have that much time; but when she does come home, she'll call. And if she can possibly make it, she'll come over, or she'll say, "Well, Grandma, can't you come over?" So we do. But like I say, we do have a very special relationship.

This close tie to one grandchild seems sufficient for Mrs. Allison; it allows her to think of herself as a person who has good relationships with her grandchildren. When she was read the story about Mrs. Smith, who was unhappy because she did not get to see the grandchildren as much as she would like, Mrs. Allison responded:

> I don't think there's really much that she can do. She should just wait and see when they can see her. . . . But I don't think she should feel, if she has a good relationship with the whole family, I don't think she should feel hurt.
>
> INTERVIEWER: *So she shouldn't take it personally.*
>
> No.
>
> INTERVIEWER: *So you would advise her to . . .*
>
> Just see them whenever she could and be content.

Mrs. Allison seems to be advising Mrs. Smith not to feel hurt as long as there are other relationships that are going well—to focus her attention on her "good relationship with the whole family" rather than on the particular relationships that aren't satisfactory. This strategy also seems to be Mrs. Allison's way of constructing her grandparental career. She maintains at least one very close relationship and accepts the lack of other close relationships—as with Christopher—as nothing personal. This allows her to think of herself as someone with a good relationship with her whole family, despite the evident unevenness of these relationships. It allows her to express considerable satisfaction about being a grandmother:

> For one thing, I know it's the continuity of the family going on. And it's somebody to love, somebody that comes to see you and that you go to see. And I think it makes you feel that you're not really getting that old.

Thus, our survey data and interviews with people like Mrs. Allison suggest that many grandparents invest more heavily in relationships with some grandchildren than with others. Particular grandchildren may be favored because they live nearby, because their parents get along better with the grandparent, or because they need help due to a family crisis. Moreover, some may just be more appealing to the grandparent because they are the first born, the last born, or the most outgoing. The emotional payoff for investment in grandchildren such as these is likely to be greater than for investment in others. Often, we suspect, a close tie to one or two grandchildren, coupled with a more distant, ritualistic relationship with the rest, may be sufficient to make grandparents satisfied with their role. They may generalize to all their grandchildren their satisfaction with their relationships with their favorites. It may not be necessary for grandparents to have equally intense ties to all grandchildren in order to feel good about being a grandparent. Equally intense relationships might even be burdensome for an older person with lots of grandchildren. Consciously or not, then, some grandparents have evolved a strategy of selective investment in which a few close ties to grandchildren suffice—in which the part substitutes adequately for the whole.

Grandparents also can adopt the strategy of creating obligations on the part of their children and grandchildren by giving gifts. Eighty-two percent of those who saw the study children in the previous twelve months reported giving money; 78 percent reported that they bought or made something for the study children. To be sure, these gifts are freely and generously offered. But, as anthropologists studying primitive societies have long noted, gift-giving is an important way of creating an obligation on the part of the recipient to give something in return.[10] The thing that is given back is not necessarily the same as what was given; grandparents, for example, do not expect to receive money from their grandchildren. But the giver and recipient both understand what sort of return is expected. In the case of grandparents and grandchildren, the expected return is visiting and affection. Mrs. Sanchez lives just outside a large northeastern city. She sees her grandchildren, who live approximately one hundred miles away, about six times a year. "I give them money when they show love," she told an interviewer. "When they come over here and I talk to them, I give them some small present, money sometimes, because I want to have their love."

The gift-giving strategy can be important when relationships are strained. Mrs. Winters explained what happened after her son and daughter-in-law divorced and the daughter-in-law refused to let her son see their children:

> And so it was a bad situation between them. But the children, I kept in touch with those children. Not by seeing them, but every Christmas and every holiday and school opening and closing I would see that they got some money.
>
> INTERVIEWER: *You would?*
>
> Yes, I'd send that money, and I even took out a Christmas saving every Christmas—a hundred-dollar savings club, and I would send it at Christmas. [Her son] didn't see them all that time, until they got old enough—you know, until they

got big enough. . . . And I told him, "You wait, one day they'll come to you."

INTERVIEWER: *And they did?*

And they did. And they meet with him now at my house, or he'll meet them at a shopping center. But he'd never go to the house.

INTERVIEWER: *Does she know that they're seeing him?*

Yeah, she knows they are seeing him. But I guess it's that she can't do anything about it, you know.

Widowed, not very well off financially, Mrs. Winters nevertheless gave money regularly because it was the only way she could maintain the grandchildren's sense of obligation to their father and paternal grandmother. When they were old enough to visit on their own (the oldest is now nineteen), the strategy paid off.

The strategies grandparents pursue also imply that the relationship may look different from the grandparents' and the grandchildren's points of view. A grandparent may be dividing her time unequally among several grandchildren. From her perspective, the total amount of effort she puts into being a grandparent may be substantial. But from the standpoint of a grandchild, especially one who is not favored with attention, her commitment may seem less consequential. We have seen, for example, that many of the grandparents who had remote or companionate relationships with the grandchildren in our study appeared to have more involved relationships with other grandchildren. Exactly how many grandparents have involved relationships with at least one of their grandchildren we cannot say from our data; but the proportion is undoubtedly higher than the 16 percent we have labeled involved on the basis of the specific grandparent-grandchild relationship we studied intensively.

Working at Kinship

In ways such as these, grandparents create their careers. By investing in some relationships at the expense of others, and by continually developing the relationships as the grandchildren grow older, grandparents do the work of grandparenting. They cannot take their status for granted or they risk becoming merely symbolic figures, emotionally distant from their grandchildren. Granted, when a person's child bears a child, he automatically becomes defined as a grandparent—but only in the narrow sense of a category in a genealogical chart. To expand that status into a meaningful, satisfying social role takes effort. That effort illustrates a basic principle of American kinship: ties with kin are recognized as meaningful only when there is an ongoing relationship, and it takes work to maintain that relationship. David Schneider, in his well-known study of American kinship, cited an extreme example: one woman he interviewed asserted that her sister wasn't a relative because she had not seen her or spoken to her in years.[11] This was an unusual case, as Schneider noted, because almost everyone counts parents, siblings, children—and grandparents, for that matter—as relatives, even if the relationship is weak. Yet in the case of grandparents, at least, this formal recognition provides only a ritualistic status.

Given the flexible nature of the American kinship system, grandparents often can choose the grandchildren to whom they pay more attention and can change loyalties as they and their grandchildren age or change places of residence. To be sure, there are constraints on their ability to choose: geographical distance, poor relationships with the middle generation, the limited number of grandchildren they may have, and so forth. Moreover, they may face competition with the other set of grandparents over the affections of some of their grandchildren. Still, grandparents, like other actors in the American kinship system, have substantial autonomy in deciding how to manage their relationships with kin. The strategies they pursue, such as selective investment, fulfill the function of allowing them to act as grandparents and to feel as though being a grandparent is an important part of their lives. These strategies may also give grandchildren a better opportunity to experience intense ties to at least one grandparent. Within families, then, all grandparent-grandchild relations are not equally close, despite the oft-repeated (and usually true) statements of grandparents that they love all their grandchildren. Instead, one often finds wide differences in the strength of the grandparent-grandchild bond, differences that appear to serve the needs of both grandparents and grandchildren to have meaningful, intense relationships with at least some members of the opposite generation.

So far we have treated American grandparents as if differences in their backgrounds and living situations were not important. This simplifying assumption has allowed us to present a broad picture of the nature of contemporary grandparenthood. But there already have been hints that geographical constraints and critical life events, such as divorce and aging, can make large differences in the experience of being a grandparent. Moreover, our training as sociologists teaches us that social class, gender, race, and other characteristics can make major differences in people's family lives. It is to these variations in American grandparenthood that we turn our attention in the next two chapters.

5

Variations

Why aren't more grandparents deeply involved in their grandchildren's lives? Why don't more grandparents play an influential role in raising their grandchildren? Why don't they help each other more? And why do some grandparents and grandchildren see each other so infrequently? The first step toward answering these questions is to examine the variations in the behavior of American grandparents. After all, previous chapters have shown that some grandparents were, in fact, deeply involved, had frequent contact, and exchanged services with their grandchildren. Clearly, there is a great deal of variation in the relationships between grandparents and grandchildren. In this chapter, we will try to determine whether this variation follows predictable patterns—whether there are certain key characteristics of grandparents and their families that allow us to predict what their relationships will be like. This search for predictable patterns should help us understand why some grandparents are more involved than others.

The Importance of Place

Let us begin with the following question: why do some grandparents and grandchildren see each other more often than others? This question, it turns out, has a simple answer: geographical distance. The effect of distance is extremely powerful—so much so that it reveals a great deal about the strength of the relationship between grandparents and grandchildren. So central is geographical proximity that grandparents who lived nearby took it for granted that they would see their grandchildren regularly. As one grandmother explained:

> I've spent quite a bit of time with my grandchildren. I've been very close to them and been able to be very close with Carole's children for the same reason: they're not that far away.

Though this explanation may seem obvious, there is no reason why grandparents and grandchildren necessarily should see each other regularly just because they live nearby. Studies show, for example, that contact among more genealogically-distant relatives—cousins, uncles, nieces, and so forth—is quite variable even when they live nearby. That Americans see their nearby parents and grandparents almost without fail suggests that strong ties continue to exist among these particular kin.

To put the size of the effect of distance in context, it must be remembered that social science research, even at its best, is not very successful in explaining variations in individual behavior. Although thousands of generalizations have been established, it is still difficult for social scientists to predict what any one person will do in a given situation. For example, although it is true that better-educated people tend to earn higher incomes, you cannot predict a particular person's income very well if you know only the level of education that person achieved. There are just too many other factors at work and too many idiosyncracies in the human personality to allow social scientists to account for much of the variation in individual behavior. Thus, most social scientists, the authors included, make a living uncovering generalizations that may be interesting and important but can only explain 10 or 20 or 30 percent of the variation in

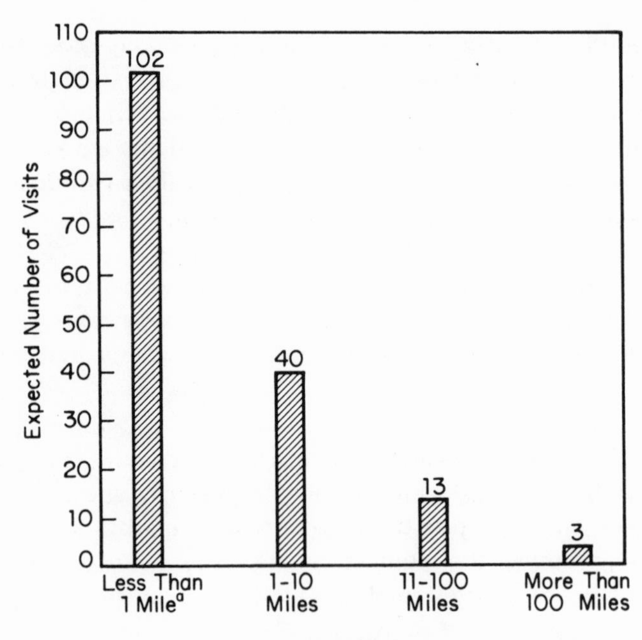

FIGURE 5–1

Expected Number of Visits per Year with the Study Child in Relation to How Far Away the Grandparent Lives

[a] Excludes grandparents who were living with their grandchildren.
SOURCE: Appendix 2.

the behavior that is studied. We were unprepared, therefore, for the strength of the relationship between how often the grandparents saw the study children (question 9) and how far apart they lived (question 1). Simply put, if you want to predict how often a grandparent will see a particular grandchild, you need to know little more than how far they live from each other. In our study, geographical distance alone accounted for 62 percent of the variance in the number of visits per year.

The overwhelming effect of distance is illustrated in figure 5–1. This figure and the following discussion are based on a multivariate statistical analysis of the frequency of visits, described in detail in appendix 2. The social-scientifically trained or statistically minded reader might wish to read this appendix. According to the calculations on which figure 5–1 is based, a grandparent whose grandchild lived more than one hundred miles away could expect only about three visits with the child per year. If the grandparent lived eleven to one hundred miles away, the expected number of visits rose to thirteen, or about one per month. If the distance was one to ten miles, the expected number was forty, or about one visit every week or two. And if the grandparent lived within one mile, the expected number was 102, or about two per week. Moreover, the effect of distance was virtually unchanged after other influences, as will be discussed later in this chapter, were taken into account.

What can we make of this powerful effect? It seems to us that the dominance of distance illustrates both the strength and the vulnerability of the grandparent-grandchild relationship. As for its strength: when grandchildren live close by, grandparents see them regularly, except under very unusual circumstances. Parents and children, with few exceptions, make sure they visit with the grandparents. The pull that grandparents exert when they live nearby shows how strong is the sense of obligation among adult children to keep in touch with their parents and their in-laws. This sense of obligation usually is overlaid with love, concern, and assistance; but even when it is unsupported by these props, it is often still honored. The uniformly high frequency of visiting among nearby kin suggests that the bond among grandparents, their adult children, and their grandchildren is still strong in the contemporary United States.

But the other side of the coin is that when adult children move away, grandparents' access to their grandchildren drops dramatically. To be sure, adult children with close ties to their parents may be less likely to move from their home towns. Job possibilities, marriage, and many other events also enter into the decision to move. Still, from the grandparents' point of view, whether or not adult children live close by involves a large element of luck. In this study, the distance between grandparents and their adult children's homes was unrelated to indicators of the grandparents' social class (such as educational attainment) or to the grandparents' marital status or health. The lack of an association between distance and social class might seem surprising, because many studies have found that better-educated persons live farther from their kin—presumably because they are more likely to move away to seek better job opportunities.[1] But a recent study of kinship ties in northern California suggests that the distance from adult children's homes to their parents' homes is an exception to this pattern,[2] and our study reinforces this conclusion. Perhaps the bond between parents and adult children is strong enough to counteract the otherwise marked spatial dispersion of kin that is found among the more educated. In any case, it would seem difficult for parents to predict or control how far away their adult children will live. When a son takes a job in another state or a daughter-in-law moves away after a divorce, the grandparent is rarely able to overcome this impediment to regular contact. Grandparents are at the mercy of their children's mobility.

Yet important exceptions to the imperatives of distance do exist. Consider the situation of Mrs. Jackson, who lives in a suburb of a large northeastern city. Her son and daughter-in-law live across the street with their two children, one of whom is the fourteen-year-old study child, Rachel. But Mrs. Jackson does not see the grandchildren very often. "I see them waving and going in and out," she told an interviewer. "That's about how much I see them, actually." Later she admitted, "I don't consider that I'm like a grandparent to Rachel. . . . You would be more friendly, I guess, with a neighbor almost." The reason for this lack of contact is that Mrs. Jackson does not get along with her daughter-in-law; and Rachel, according to Mrs. Jackson, "is very close to her mother" and to her mother's side of the family. Mrs. Jackson is not sure why she and Rachel's mother have such a poor relationship, nor does she know how to improve it. "I wish there were a closer relationship," she said, "and I would go more than my half to do it, but I don't know how to do it."

Mrs. Jackson's plight shows that the quality of the relationship between a grandparent and her daughter or daughter-in-law can be a major determinant of the frequency of contact. Past studies have suggested that women, more than men, do the work of maintaining intergenerational solidarity,[3] and our preliminary interviews with grandparents also suggested as much. Because daughters and daughters-in-law often serve as gatekeepers, grandparents who do not get along with them may find it more difficult to see their grandchildren. In most families, the relationship between grandparents and their daughters or daughters-in-law is not a problem: three-fourths of the grandparents in the study said their relationship with the study child's mother was "extremely close" or "quite close." But in a minority of families, such as the Jacksons, tensions between mothers and grandparents are a barrier to frequent contact.

Our statistical analyses in appendix 2 bear this out. Even after the effects of distance and other factors were accounted for, grandparents who reported that their relationship with the study child's mother was "extremely close" saw their grandchildren about twice as often as those who said "fairly close" or "not very close." That this large differential was still apparent despite the enormous influence of distance suggests that mothers really are important bridges between the generations. We say mothers, rather than parents, because the role of fathers in maintaining intergenerational contact appears to be much smaller. Further analyses showed that once the quality of the relationship between a grandparent and the study child's mother was taken into account, the quality of the grandparent's relationship with the study child's father had little effect on how often the grandparent saw the grandchild. The lack of importance of relationships with fathers even held true for paternal grandparents.[4] Grandparents who want frequent contact with their grandchildren need to get along well with their daughters or daughters-in-law, but their relationships with their sons or sons-in-law make much less difference. Mothers are the gatekeepers, but they do not appear to play favorites: the maternal grandparents in the study did not see their grandchildren much more often than the paternal grandparents did. This finding is consistent with an unwritten rule of our kinship system, namely, that both sides of the family have an equal right to see the grandchildren. Still, we must be cautious in reporting this seeming equality, because the mothers who provided the names of the grandparents in the study may have withheld the names of in-laws whom they saw infrequently.

When we were conducting preliminary interviews, we sensed that certain families had more of a family consciousness than others. They seemed to place more emphasis on family lore, rituals, and events, to have more of a family spirit, and to care more about keeping the family together. This is an elusive, intangible characteristic that is difficult to measure. Nevertheless, we asked several questions in the survey designed to reflect differences in family consciousness. Our hypothesis was that grandparents in families with a great deal of family spirit would see more of their grandchildren, other things being equal. Three questions, in particular, seemed to tap this dimension.[5] Prefaced by the phrase "In your family," they were as follows:

1. Are there special family recipes or dishes?
2. Are there family jokes, common expressions, or songs?
3. Are there ritual or special events that bring the family together?

Mrs. De Lorenzo, who lives in a large metropolitan area in the Midwest, answered affirmatively to all three questions. She sees her sixteen-year-old grandson about once a week, somewhat more often than would be expected, given that they live about thirty miles apart. In a follow-up interview, she was asked about family rituals and family consciousness:

INTERVIEWER: *When you say you have a close family, what is that?*

Well, we're always together for holidays and everything—we're always together. We always keep in touch, anyway, and see each other quite often.

INTERVIEWER: *You said in your interview that you have some family recipes and things like that, family expressions. Can you tell me some of those?*

Well, there's a lot of old family recipes that we have, and I try to hand them down to the grandchildren.

INTERVIEWER: *So they'll say, "This is Grandma's."*

That's right. And I make them things like afghans, crocheted doilies, and stuff. I always say, now remember, you keep that, because one of these days I'm not going to be here, and at least you'll have something and you will think about me.

INTERVIEWER: *So they really value their relationship with you?*

They tell me not to say I'm getting old and I'm going to die pretty soon. They say, "Oh, Nonna"—that's grandmother in Italian—"don't talk like that." Well, it's true, I'm not going to be here forever.

INTERVIEWER: *Do you have any other Italian customs, words, recipes, poems, or anything like that?*

[Laughter] I used to sing to them when they were little—Italian songs that my grandfather used to sing to me. They are interested in everything to do with Italy.

INTERVIEWER: *If some neighbor dropped in and you were talking and they said, "What can I do to make my family feel closer to each other?" what would you tell them?*

Well, I'd tell them that you have to keep in touch and just

tell them how much you love them and everything. You have to show love to get love.

In the study as a whole, grandparents who reported more family rituals did see their grandchildren more often. For our statistical analyses of the frequency of visits, we excluded the third question (special events that bring the family together, p. 112) since it seemed too similar to the behavior we were trying to predict. Even so, grandparents who said that they had "special family recipes or dishes" and "family jokes, common expressions, or songs" saw their grandchildren 50 percent more often than those who had none of the above—even after controlling for distance and the quality of the mother-grandparent relationship.[6] That this brief, specific, two-question scale makes a substantial difference despite the constraints of distance and personal relationships suggests that family consciousness is a real and significant component of family life. Unfortunately, we cannot say much about who these families are and why they have such a familistic orientation. Further tabulations of the survey data showed only that rituals were more common among grandparents who were somewhat better educated, slightly older, white, and puzzlingly, on the paternal side.

Two other determinants of the frequency of contact deserve mention, although their effects are somewhat smaller.[7] First, grandparents who provided some financial support to the child's parents in the previous twelve months saw the grandchildren 32 percent more often. Financial support may indicate a broader pattern of assistance that could lead to more frequent contact. But it is possible that the causal connection goes the other way: grandparents who see their grandchildren more often may be more inclined to give support. Second, grandparents who lived in rural areas (places with a population of less than 2,500) saw their grandchildren more often, other things being equal. We cannot say for sure why rural grandparents reported more contact. This pattern may reflect different values among rural families, values that lead to more social contact or perhaps a greater amount of mutual assistance. Among urban grandparents, those in larger areas (places with populations of 100,000 or more) saw their grandchildren somewhat more often than those in smaller areas (2,500 to 99,999 population). In larger urban areas, superior transportation may make visiting easier and quicker.

Do grandparents who see their grandchildren infrequently compensate by talking to them more often on the telephone? No, to judge from what the grandparents in our study told us about their telephone conversations with their grandchildren. The same grandparents who see the study children least often —those who live farthest away—also talk to them least often on the telephone.[8] A telephone call may be the next best thing to being there, as the telephone company used to remind us, but it cannot compensate for living far away. Rather, telephone calls are more frequent among those who visit more often—suggesting that most calls serve the purpose of furthering routine, regular interaction rather than renewing distant relationships. Grandparents call their children and grandchildren more often to ask for a favor or to report what happened yesterday than to wish them a happy birthday.

On the whole, though, the frequency of telephone contact between grandparents and grandchildren is much less predictable than is the frequency of visits, as our analysis in appendix 2 demonstrates. The list of predictors discussed previously (distance, quality of relationship with study child's mother, familism scale, and so forth—for the complete list see appendix 2) accounted for 69 percent of the variance in visiting frequency; the same list accounted for just 20 percent of the variance in the frequency of telephone calls. This lower predictive power implies that characteristics such as distance and the quality of relationships have less influence on the frequency of telephone calls than on the frequency of visiting. In general, the determinants of frequent telephone contact were the same as the determinants of frequent visiting. The major difference was that maternal grandparents talk to their grandchildren on the telephone more frequently than do paternal grandparents—about 50 percent more, according to our analyses.[9] We suspect this is because mothers do much of the telephoning, and mothers tend to call their parents more than their in-laws. As a result, paternal grandparents talk less to their grandchildren by telephone, although they see them just as often as maternal grandparents do.

In sum, there is no mystery about which grandparents see their grandchildren frequently, no need to appeal to the unpredictability of human behavior or the limits of social science. The frequency of visits between the grandparents and grandchildren in this study was highly predictable: by asking about a few determinants, we could account for 69 percent of the variance in the number of times they had seen each other in the previous twelve months. The major, dominating influence was geographical distance, which alone accounted for 62 percent of the variance. To be sure, it mattered how well the grandparent got along with the child's mother, to what degree a familistic orientation existed, whether or not the grandparent was providing financial support, and whether the grandparent lived in an urban or rural area. But the effect of distance was overwhelming.

As an illustration, consider two hypothetical paternal grandmothers. The first has an "extremely close" relationship with her daughter-in-law, knows family recipes and songs, provides financial support, resides on a farm—but lives more than one hundred miles away from her grandchild. The second is "not very close" to her daughter-in-law, has no recipes, jokes, songs, or other rituals, does not provide financial support, resides in a small city—but lives within ten miles of her grandchild. The latter grandparent, according to the statistical estimates presented in appendix 2, would see her grandchild about twice as often as the former, despite her strained kinship ties and lack of financial support and family ritual. If asked to name the three most important determinants of the frequency of contact between grandparents and grandchildren, we would have to reply: distance, distance, and distance.

To be sure, geographical distance also influences our relationships with other relatives, but not to the same degree. In a detailed, national study of the patterns of contact between married men and their relatives, Sheila Klatzky found that geographical distance had much more of an effect on relationships between genealogically close kin such as fathers and sons, than on genealogically distant kin such as uncles and nephews. The men in her sample who lived near their fathers saw them often, and men who lived farther away saw their fathers much less often. In contrast, uncles and nephews did not see each other much more often if they lived in the same neighborhood than if they lived fifty miles apart.[10] These results suggest that geographical distance makes the most difference in relationships that we feel obligated to maintain and, furthermore, that these feelings of obligation are much stronger for relationships of direct descent—father, son, grandson—than for lateral relationships. Since the obligation to visit one's uncles is much less strongly felt than the obligation to visit one's parents, an adult is freer to avoid frequent contact with his uncles (or cousins), even if they live nearby. In a kinship system such as ours where people trace kin through both their father's and mother's lines, the pool of genealogically distant kin is large. Americans have the option of choosing which of these potential links to turn into real relationships. The direct link between parents and children, however, is different; strong social norms still ensure that grandparents will see their adult children—and the grandchildren who live with their children—often, except where long distances or other exceptional circumstances exist.

Before leaving this topic, let us mention the many characteristics of the grandparents that were *not* associated with frequency of contact: social class (as measured by the educational attainment and family income of the grandparents), sex, race, age, health, marital status, labor force status, and (for visits but not telephone calls) whether the grandparent was on the maternal or paternal side. (But differences between maternal and paternal grandparents may have been understated somewhat by our sampling procedure, as is discussed in appendix 1.) These negative results attest once again to the strength of the ties between grandparents and their adult children and grandchildren. Regardless of their background or personal activities, grandparents see their children and grandchildren often when they live nearby, especially if they have a decent relationship with mothers in the middle generation.

Yet the findings presented so far pertain only to the quantity of contact. The quality of that contact is another matter. Not all visits are the same, not all are as meaningful or as helpful or as memorable as others. As we have seen, there is great variation among grandparents—even among those who see their grandchildren regularly—in the style of their interaction with their grandchildren. Why, we will ask next, are some grandparents more involved with their grandchildren than others?

Who Is "Involved"?

The reader will remember that in chapter 3 we defined involved relationships on the basis of two sets of questions in the survey. The first set—of five questions—measured the extent to which grandparents were able to act toward their grandchildren in ways usually reserved for parents: disciplining, correcting, being consulted about major decisions, and so forth. The second set—of four questions—measured the exchange of services: whether the grandparents and grandchildren helped each other with errands or chores, or whether they bought or made things for each other. For each set of questions, the number of positive responses was summed to form a scale score. Scores on the parentlike behavior scale ranged from 0 (no positive responses) to 5 (all positive responses), with an average score of 2.3. Scores on the exchange scale ranged from 0 to 4, with an average of 1.8. In the second section of appendix 2, we present a statistical analysis of the variation in the two scale scores among the grandparents.

The results show that for both scales, the most important predictor (by far) of who scores higher is frequency of visiting. Even after a number of other factors (as will be discussed) were statistically controlled, its effects were quite large. Other things being equal, a grandparent who saw the study child almost every day during the previous twelve months had an average score of nearly 4 on the scale of parentlike behavior; and a grandparent who saw the study child once or twice a month had an average score of 2.[11] This is a large difference on a five-item scale. For example, the latter grandparent might have said merely that she gave the child advice and discussed his problems; but the former may have done all that and, in addition, disciplined the child and been consulted before important decisions about the child were made. The effect of frequent visits on the exchange of services was equally impressive.[12] The scale scores dropped rapidly as frequency of visits diminished: grandparents who saw the study children almost every day (some of these grandparents lived with the study children) had much higher scores; grandparents who saw the study children a few times a week scored lower but still considerably above average on both scales. This pattern suggests that quite regular, almost daily, contact is necessary if a grandparent is to play an active role in childrearing and exchanging services. And as we have seen, frequent contact is itself highly dependent on living nearby. No other predictor approaches the magnitude of the effect of frequent contact.

Several other factors, however, do make a difference. Younger grandparents reported more parentlike behavior and exchanges of services. We suspect that some older grandparents lack the physical energy to engage in an involved relationship. Black grandparents were considerably more likely to act like parents, even after statistical controls for family income, age, frequency of contact, divorce and separation, and relationships with the study child's mother. Even after these other factors were taken into account, black grandparents scored, on average, more than one point higher on the scale of parentlike behavior. More will be said about the role of black grandparents shortly. In addition, grandparents whose grandchildren's parents have separated or divorced scored higher on the parentlike behavior scale, other things being equal. (But note that most of the grandparents in this situation are maternal grandparents because of the limitations of our sampling procedure. In chapter 6 we will describe the very different effect that marital dissolution has on maternal versus paternal grandparents.) Two other factors make a difference for the exchange of services, other things being equal. Grandparents who report more family rituals, presumably an indicator of a stronger orientation toward family life, exchange more services, as do those who have a close relationship with the study child's mother.

A few other factors were notable for their lack of effect. Family income did not have a significant effect, after statistical controls for other variables, on either scale. Further analyses showed that neither did the grandparents' educational attainment. These negative results are consistent with the general lack of social class differences in the study: recall that social class differences did not affect the frequency of visiting. Nor did self-reported health (question 117) have an effect, although age, as noted, was an important predictor. Other variables that might have been expected to influence the scale scores but did not: religiosity (question 107), whether the grandparent provided financial support to the parent (question 125), and the sex of the grandparent.

Why, then, didn't more grandparents have involved relationships with the study children? Because families must have a special set of characteristics in order to have involved relationships, and most lack one or more of these characteristics. First, and above all, grandparents must see their grandchildren very frequently. That, in turn, requires that the grandparent and grandchild live near each other—within a few miles or, better yet, within a few blocks. But given the great amount of geographical mobility in American society, it is not possible for most grandparents to live nearby. Second, involved relationships are more common when a disruptive life event—such as a divorce in the middle generation—creates a need for assistance from the grandparents. Unfortunately, many more families are experiencing divorce than was the case a generation ago; but even when a divorce occurs, the grandparent may not be near enough to play an active, supportive role. Moreover, younger grandparents (who tend to be more vigorous) are able to undertake activities on behalf of their grandchildren. And cultural values (as in black families and as measured by questions about family songs, rituals, recipes, and so forth) create a climate that is conducive to greater involvement. Overall, a set of quite special circumstances must all be present for grandparents to play a major role in rearing and supporting their grandchildren. For most grandparents and grandchildren, one or more of the critical ingredients for an involved relationship is absent; as a result, this style is relatively uncommon.

These findings suggest to us that the proportion of grandparents and grandchildren with involved relationships will not increase much unless there are major changes in the American kinship system. This point deserves further elaboration, and we will return to it in the final chapter. There is, however, one social trend that is bringing grandparents and grandchildren closer together—at least along the maternal line. Ironically, it is a trend no one would encourage: the great increase in divorce. We will discuss its effects in detail in the next chapter.

But what about the types of variation that reflect basic social characteristics such as gender, ethnicity, and class? These are the staples of sociological analysis; social scientists have shown again and again how these basic characteristics exert powerful influences over individuals' lives. So far, our analyses have not shown much of an effect of these forces, with the exception of race. Can this be true? They deserve a closer look.

Gender

A long tradition of sociological research on the family suggests that the roles of men and women are distinct. Ties between mothers and daughters are said to be the strongest kin links; married daughters and their mothers supposedly do much of the work of keeping families in touch. Women, in other words, are said to be the "kin-keepers" or "lineage bridges" whose efforts hold kin groups together.[13] Moreover, an influential line of research holds that men specialize in task-oriented, "instrumental" family roles, such as working at a job and keeping the house and car in shape, while wives specialize in nurturant, emotion-laden, "expressive" roles, such as rearing children and comforting adults.[14] To be sure, much of this research dates from the 1950s and early 1960s, a period of family life that now seems like ancient history: most wives did not work outside the home, birth rates were high, divorce rates relatively low, and so forth. But that was the period when today's grandparents were marrying and having children. Moreover, despite the great changes in family life, the household division of labor between husbands and wives continues to be unequal. There is reason to expect, therefore, that the roles of grandmother and grandfather may differ.

Good information on gender differences in grandparenthood is, however, rare. The very limited amount of information that does exist suggests that the differences between grandfathers and grandmothers occur along traditional, sex-typed lines. Grandfathers, some observers have argued, are more instrumental in their behavior toward their grandchildren, concentrating more on shared tasks and activities. Grandmothers do more of the kin-keeping and have warmer, more expressive relationships with their grandchildren.[15] Gunhild Hagestad has argued that grandfathers distinguish much more between grandsons and granddaughters than do grandmothers.[16] All agree, though, that not much is known.[17]

Our study confirms that differences do exist between grandmothers and grandfathers and that these differences do break down along traditional lines. Yet the differences are not overwhelming: recall that, other things being equal, grandfathers and grandmothers saw the study children a similar amount, took on a parentlike role about as often, and exchanged services about as often. But on closer examination, the data on exchanges of services support Hagestad's assertion that the sex of the grandchild matters more for grandfathers than for grandmothers. Consider the percentages of grandparents in our survey who scored high (3 or 4) on the exchange scale, broken down by the sex of the grandparent and of the study child:

When the grandparent was:	and the grandchild was:	the percent who scored high was:
male	male	64
male	female	30
female	male	34
female	female	36

Among all grandfather-grandson pairs, then, about two-thirds exchanged substantial amounts of help with chores, errands, or activities. Among all other pairs (grandfather-granddaughter, grandmother-grandson, grandmother-granddaughter), the proportion was only about one-third. This pattern held even after statistical controls for the frequency of visiting and the age of the grandparent.[18]

The distinctive, instrumental nature of the relationship between grandfathers and grandsons was apparent in some of our follow-up interviews. The reader will recall Mr. Sampson, the "involved" grandparent quoted extensively in chapter 3, who used to take trips with his grandsons, fix radios with them, put up window screens together and, in general, engage in "a terrible lot of activity." Or consider Mr. and Mrs. Harris, who had just returned from a visit with their son and three grandsons, including the study child, Roy, when they were interviewed. When Mr. Harris was asked what he did with Roy when they were together, he replied:

> Well, it wasn't too much of anything other than working out in the yard and helping him a little bit with his car that his older brother wrecked on him. And things like that, that's all. . . . And they [Roy and his brothers] were helping me do some painting for my son down there, all that kind of stuff.

Asked the same question, Mrs. Harris replied:

> Oh, sit and talk, more or less, about his future more than anything. And what he's interested in doing. That's about the extent of it. Teenagers, I don't think, feel that older people have as much knowledge about, oh, their friends and things like that.

Thus, while her husband busied himself with activities, Mrs. Harris talked with her grandson about hopes and feelings. His relationship with Roy was more instrumental, hers more expressive. She was engaged in trying to bridge the gap between the generations, an effort that Hagestad claims grandmothers make more successfully than grandfathers.[19]

It may be that grandfathers, at least this generation, can express themselves better through actions than through words. One grandmother commented:

> My husband is not—well, he's affectionate in one way. But not as much as I am. I think he feels that you don't overin-

dulge in affection when it comes to the children, you know; this is the feeling I have. In other words, he's more the exact type of person.

The "exact" type of person, who tends to be a male type of person, does not express his feelings as easily or openly. Mrs. Harris had no difficulty expressing her feelings to an interviewer. When asked how frequently she thought about her grandchildren, she answered:

All the time, all the time. Particularly with a lot of them driving now. That worries me a lot. More than anything, I worry about them driving cars, accidents. I don't know if I could stand it.
INTERVIEWER: *So for you, being a grandmother is a day-to-day thing.*
Yes, right. Absolutely. They're my whole life, actually, all the kids—the grandchildren.

But her husband had great difficulty with a similar line of questioning:

INTERVIEWER: *What do you think it's meant for you, for your life, to be a grandfather?*
What do I think it's meant to me?
INTERVIEWER: *What do you think it's meant for you? And what do you think it means to you now, to your life, to have been a grandfather?*
I don't get that one.
INTERVIEWER: *Well, has it been important to you? Do you ever think about what it is to be a grandfather?*
Well, I've thought about it. The only way I've thought about it is I have to sleep with grandmother [laughter]!
INTERVIEWER: *So you don't really think much about being a grandfather. Do you just sort of take it for granted?*
I just take it for granted, and I'm glad, I'm happy that I am

a grandfather. Because the eight [grand-] children that I have, we've never had a problem with them, you know. And that means a lot. I'm happy that I'm a grandfather.

After much initial confusion and some lame joking, the best Mr. Harris can come up with is that he is happy that he is a grandfather. Emotional expression is not easy for him.

This reluctance or inability to express feelings on the part of some males may help explain why women do the kin-keeping. Even though the grandfathers in our sample see their grandchildren as often as grandmothers do, it is often the grandmothers and their daughters (or daughters-in-law) who make the arrangements. Here is a conversation with a grandmother from a northeastern city:

INTERVIEWER: *When you see [your granddaughter] is your husband usually present?*

Yeah, I mean most of the time. Because, after all, he's retired; we go together, you know.

INTERVIEWER: *And you said you usually make the plans with [your daughter]?*

Yes, yes.

INTERVIEWER: *So your husband isn't really involved in the plans initially?*

Well, I'll say, "Is such-and-such a time suitable for you," does it fit in, you know. If it is, he'll go along with it; if it isn't, well, we just have to postpone it for a couple of days or a more opportune time.

INTERVIEWER: *How close do you think your husband is to [your granddaughter]?*

He's probably as close as I am. Only he's not quite so verbal, let's say, about it.

In fact, grandfathers in the survey reported themselves to be as close to their grandchildren as did grandmothers. Yet this similarity may be due in part to the difficulty we encountered in completing interviews with grandfathers. As discussed in chapter 1, we had designed our sampling procedure to yield one hundred interviews with grandfathers but obtained only sixty-four. Granted, grandfathers were more likely to be working outside the house or to be older (and therefore in poorer health) than grandmothers, so they were harder to interview. But we believe that our difficulties also reflect the greater reluctance of grandfathers to talk about their family lives. At any rate, the grandfathers who did cooperate may have been those with closer relationships with their grandchildren.

These gender differences can be summed up in one principle: grandfathers are to grandmothers as fathers are to mothers or, indeed, as men are to women in our society. The differences that have been described fit the general pattern of gender differences in family and personal life. We suspect that the gender differences between grandparents are not as great as those between parents because neither grandfathers nor grandmothers typically are involved in day-to-day childrearing. The differences we did observe also may reflect the particular values of adults who married and had children during the familistic 1950s. Perhaps grandfathers and grandmothers will become more alike when the parents of the 1970s and 1980s reach grandparenthood. But even among younger couples today, gender roles still exist. It is unlikely, then, that the differences between grandfathers and grandmothers will fade away soon.

Race

As noted previously, we found a substantial difference between black grandparents and white grandparents. (Our samples of Hispanic and Asian-American grandparents were too small to enable conclusions to be drawn.) Black grandparents were much more likely to take on a parentlike role with their grandchildren. On some of the survey questions measuring parentlike behavior, the differences were very large: 87 percent of black grandparents said they correct the study child's behavior "often" or "sometimes" when they see him or her doing something they disapprove of, compared to only 43 percent of white grandparents. Seventy-one percent had disciplined the grandchild over the previous twelve months, compared to 38 percent of whites.[20] Recall the black grandmother at the senior citizen center who proudly said that her grandchildren call her "sergeant" and who checked up on them at school. Among the grandparents at the center, behavior such as this was the rule. Another grandmother at the center, Mrs. Lewis, told the interviewers:

> I don't even know why I'm in here, because my story is so uneventful. I have a daughter who lives with me with one child. But I wouldn't say I was raising this child. I bought a house with four apartments to be together, since my husband died and she was divorced, so to help raise this child, to be closer to her. I've got this house, and she's on the third floor and I'm on the second floor. But this child, when he first came to us, he was a small child. He's fourteen now. Now that he is fourteen years old, I'm going to tell you right now, he is no angel. You hear me? He calls me the worst grandmother in the world, the meanest one. Because this child doesn't have a father there with us, and I was raised in a family without the father's image. So that means I've got to be a little stronger, we've got to be very strong, with what's out there in this world now, and I've got to keep him—he can get around his mother, but he can't get around me so well.

Mrs. Lewis's sense that she has got to be a little stronger has led her to take on a pseudoparental role, one that she takes great pride in. She described in detail how she watches out for her grandson, fixes dinner for him when his mother is late returning from work, and makes sure he cleans his room. Her repeated denials that she was raising the child were not taken seriously by the others in the room—nor, it seems, were they meant to be. Her friend, the "sergeant," interrupted at one point:

> You might not call yourself basically rearing him, no, because he does have a mother with him—your daughter. But you're correcting and keeping him up the right, straight, and narrow path, because you're there when the mother's out to work, and you see, like his room. . . . When a child's in an adolescent stage with things like they are in the street today, they need some strict supervision. If you don't give them ground rules to work on, and some supervision, and some time and things to do things. . . .

These grandparents saw themselves as protectors of the family, bulwarks against the forces of separation, divorce, drugs, crime —all of the ills low-income black youth can fall prey to. Mrs. Lewis continued:

> And I tell you, I don't know in the white families, but I'm going to tell you in the black families, so many kids have to come up just with the mother image. They don't have a father at the home, and they're divorced or separated or whatever, but so many black children come up with nobody but the mother image. My children were fortunate, they had a father and a mother there. My husband's been dead just about five years, so he was with me to raise these children.
>
> INTERVIEWER: *What difference do you think it makes that the children are growing up with just the mother image, as you said?*
>
> I think they don't get a father in a home, whether he's got authority or not. A man's image would make these children understand something. That's the reason I think I've got to be a little stronger than my daughter with this boy at this age.

With all these pills and one thing or another out there now, you sometimes got to take it.

INTERVIEWER: *Sounds like in some ways you've got to be both, you've got to be something like the father. . . .*

You do. Because my mother was like that. See, my father and mother separated when I was seven years old. And with four children, she had to be a father and mother.

This type of pseudoparental behavior gives grandparents such as Mrs. Lewis a sense of pride and accomplishment, a feeling that they are central to their families' struggles to get by. But this instrumental role is something black grandparents do, as Mrs. Lewis's remarks make clear, because they have to. It is not a style of grandparenting freely chosen; rather, it is a style adopted under duress. It is rooted in past and present experiences with hard times. It reflects the continuing instabilities that separation, divorce, unemployment, illness, and death can bring to black family life. At the turn of the century, when family life was less stable and predictable for white and black families, this may have been a much more common way for American grandparents to behave. Black grandparents may represent the last holdout of a form of grandparenting in which companionship can sometimes take a back seat to authority.

This is not to say that black grandparents are not also loving companions. One need only remember Mrs. Winters's deeply felt remarks about how grandparents have "love to spare." Or the black grandmother, quoted in chapter 1, who said that being a grandparent is different than in the past: "Grandparents are more indulgent because they don't want to lose their grandchildren's love." Moreover, all of the quotations so far have come from grandmothers. Indeed, the two grandfathers at the senior center had little to say. "Grandfathers are a little softer," one grandmother told us, to general approval from the others in the room.

The difference is one of degree. Forty-four percent of the black grandparents in the survey reported that they had lived with the study children for three months or more, compared to 18 percent among whites. Black grandparents were more likely to discipline their grandchildren, as has been noted, but less likely to joke or kid with them (78 percent versus 92 percent among whites). Twenty-seven percent had "involved" relationships with the study children, compared to 15 percent of whites.

The greater instrumental role for grandmothers is consistent with reports of female-centered kin networks among low-income blacks.[21] These networks provide a way for families to survive the hardships of poverty by sharing resources. Yet it is important to note that our statistical analyses showed that the racial difference did not occur only among the poor. As noted earlier, black grandparents tended to score higher on the scale of parentlike behavior even after statistical controls for family income, marital dissolution, and other factors. The distinctive pattern we observed among black grandparents—or at least among black grandmothers—is not just a function of poverty or family structure. Developed during centuries of adversity, it has now become part of the cultural repertoire of black families—even middle-class black families. A study by Harriette P. McAdoo of middle-income blacks in the Washington, D.C., area revealed a substantial amount of assistance from other kin.[22] Many of the upwardly mobile parents stated that they would not have succeeded without the support of kin. Mothers reported that childcare was the most important type of help they received from kin, and fathers cited financial help as most important. Where middle-class status is still precarious, as in many black families, extensive exchanges among kin, including grandparents, can be a critical source of support.

The Relative Unimportance of Class

As mentioned earlier, a number of studies have found strong differences in family life according to social class position, so we were on the lookout for class differences in grandparenthood.

We did not find any. To review, better-educated and higher-income grandparents did not see their grandchildren less often, as might have been expected. Moreover, education and family income made no difference, other things being equal, in how much of a parentlike role grandparents took on or in the degree to which services were exchanged. In a subsequent chapter, we will report that higher-income grandparents were no more or less successful in passing on values to their grandchildren than were lower-income grandparents. Why this absence of class-related differences? We can think of three plausible explanations.

1. *The exception to the rule.* Perhaps the well-known tendency of better-educated people to live farther from their kin does not apply to parents and children. As we noted, a study of kinship ties in northern California found that better-educated persons, as expected, lived farther from relatives, with the exception of the distance to their parents. Our results fit this pattern. If the distance between parents and children truly is an exception to the general pattern, it would suggest that the ties between older parents and adult children are far stronger than all other ties with relatives living elsewhere—so much so that even potentially mobile, well-educated children make an effort to remain near their parents. This would be evidence of the continuing strength of the grandparent-parent-grandchild bond.

2. *The narrowing of class differences.* In 1924 Robert and Helen Lynd began their well-known study of Muncie, Indiana, the results of which were published in *Middletown* and *Middletown in Transition.*[23] The Lynds found strong differences between "business-class" families, with full-time housewives, spacious homes, modern kitchens, and domestic help, and "working-class" families, with smaller homes, coal- or wood-burning stoves, outdoor privies, and wives who often worked outside the home. Fifty years later, Theodore Caplow and his associates restudied Muncie to see what changes had occurred. They concluded:

> Most of the differences that the Lynds observed between business-class families and working-class families half a century ago have by now eroded away. Working-class people play golf and tennis, travel in Europe for pleasure, and send their children to college. Business-class people do their own laundry and mow their own lawns. Business-class wives with children at home are as likely to hold full-time jobs as working-class wives. There is more decorum in working-class churches and more fervor in the business-class churches than there used to be, and many congregations are thoroughly mixed.[24]

They attribute this convergence to the slowing of the pace of modernization, which has allowed working-class families to catch up in terms of lifestyle, if not income.

In general, the trends in American family patterns have moved in the same direction for all major subgroups in the population over the past few decades. Fertility rose among all groups in the 1950s and has fallen pervasively since; marital disruption appears to have increased throughout the population between the early 1960s and the mid-1970s; and the rise in married women's labor force participation has been widespread.[25] Some sharp differences between blacks and whites still exist—such as in out-of-wedlock childbearing. But at least within the white population, one might advance the hypothesis that social class differences in family structure, childrearing, and intergenerational relations have become smaller, and in some cases negligible, in recent decades.

3. *The dominance of life events.* Most grandparents, as has been demonstrated, are only marginally involved in the rearing of their grandchildren. In most families, grandparents serve as a latent source of support, ready to step in when needed. Under these circumstances, one might expect that any existing class differences would be overwhelmed by life events. A long-distance move deprives middle-class and lower-class grandparents alike of the opportunity to see their grandchildren regu-

larly. Even in middle-class families, a divorced child can require substantial help from parents. To be sure, lower-class and working-class families are more vulnerable to crises such as divorce or unemployment. But divorce has become common even in the middle class. Moreover, we have seen how the geographical distance between parents and adult children appears to be unrelated to education and income. If one wants to predict the kind of relationship a grandmother has with her grandchild, one would be much better advised to ask how far apart they live or whether the parents are divorced or whether the father is stationed overseas than to ask about her education or her income. In other words, even if class makes a difference, life events make much more of a difference.

We believe that there is some truth to all three of these explanations. But until more is known, they must remain speculative. In any case, the absence of class differences is a clear, and somewhat surprising, finding from our study.

Ethnicity

The life events explanation might also account for the apparent lack of substantial ethnic differences among the white grandparents we interviewed. This conclusion must remain tentative, for our national sample was too small to statistically compare grandparents from the many ethnic groups. But from the start of our project, when we talked with grandparents at senior citizen centers in Polish-American and Jewish neighborhoods, we found that ethnic differences among white grandparents seemed modest. This impression was somewhat surprising given the great amount of folk wisdom—and some empirical research—about ethnic differences in family life. There are the familiar stereotypes of the warm Jewish grandmother making matzoh ball soup, the warm Italian grandmother cooking tomato sauce, the warm Irish grandmother dishing out corned beef and cabbage, and so forth. Each group thinks of its grandparents as loving and warm, and each, it turns out, is right—but only because, as we argued in chapter 3, most American grandparents have these characteristics.

The point is that when we examined the attitudes and behaviors of the white grandparents we interviewed in person before and after our national telephone survey, we found little difference according to religion or nationality. Neither the Poles nor the Jews could violate the norm of noninterference except under special circumstances such as a divorce. Both groups spoke of love and companionship as central to their definition of the grandparental role. The phrases that they used differed somewhat and the ritual occasions on which their families gathered varied, of course; but on a day-to-day level their relationships with their grandchildren seemed similar. As with social class, we believe that strong ethnic differences failed to emerge because most grandparents, whatever their ethnic group, are not very involved on a day-to-day level. Therefore, the constraints of distance or the exigencies of divorce define the nature of the relationship much more than the cultural variations among American ethnic groups. The previously noted exception—the distinctiveness of black grandparents—proves the rule, for black grandparents, by and large, are much more heavily involved in their grandchildren's everyday lives. Where a high degree of involvement is lacking, ethnic variations appear to be small.

There is one life event—divorce—that has increased in frequency to the point where few extended families are untouched by it. A divorce in the middle generation has the potential to restructure the relationships among grandparents, their children, and their grandchildren. When coupled with prevailing custody and visitation patterns, it also has the potential to alter the structure of American kinship. It is to the effects of divorce that we now turn.

6

A Special Case: Grandparents and Divorce

It would seem obvious that when parents divorce, grandparents are affected too. Yet the effect of the recent increases in divorce and remarriage on kinship practices and intergenerational relations remains an intriguing but largely unexplored topic. What little we do know comes from a few small-scale studies which suggest that in the aftermath of a divorce, the ties between a divorced woman and her parents are generally strengthened, but relations with her former in-laws drop off sharply.[1] This finding has led several investigators to conclude that divorce contracts the kinship network and weakens the bonds between generations.

But this conclusion is too simplistic. A woman may stop seeing her former husband's family but it does not follow that her children do. For example, Furstenberg and Graham Spanier reported that children in divorced families saw their fathers' relatives, including their paternal grandparents, much more often than did their mothers. And Colleen Johnson, in a study of grandmothers in the San Francisco Bay Area, reported that some paternal grandmothers established working relationships with their ex-daughters-in-law, providing assistance in order to preserve their relationships with their grandchildren.[2] These findings suggest that grandparent-grandchild relations sometimes persist along the noncustodial side after divorce. It also is possible that weakened intergenerational ties on the noncustodial side (usually the father's side) could be more than compensated for by strengthened intergenerational ties on the custodial side. Moreover, a parental divorce is often followed by a remarriage, which has the potential to expand the number of intergenerational ties.

In this chapter, we will show how divorce creates both opportunities and dilemmas for grandparents. The opportunities arise from the need parents and children have for assistance after a divorce. Grandparents, especially those on the custodial side, can maintain or even deepen their relationships with children and grandchildren by providing material assistance, a place to live, help in childrearing, guidance, or advice. The dilemmas arise from the constraints imposed on grandparents, particularly on the noncustodial side, by the actions of the middle generation. The divorcing parents, for instance, will often wish to restrict their relations with one another as much as possible. Since the custodial parent frequently controls access to the grandchildren, the grandparents' access may be restricted unless cordial relations with the daughter- or son-in-law are maintained. Consequently, the grandparents on the noncustodial side, as Johnson's study demonstrated, may have a stake in preserving relations with their former child-in-law, even at the risk of offending their child, if they wish to maintain ties with their grandchildren.

More generally, the study of intergenerational relations after divorce provides us with an opportunity to see the ways in which the nature of American kinship is currently being played out. At current rates, about two-fifths of all children will experience a parental divorce.[3] During family crises such as divorce, adults turn to their kin living elsewhere—particularly to their parents—for help.[4] Because latent kinship ties are likely to be activated, the events surrounding and following divorce provide a special lens through which we can view kinship and make its meaning clearer. For example, later in this chapter we will compare grandparent-grandchild relations in disrupted versus nondisrupted families, with the comparisons drawn separately for maternal and paternal grandparents. It is well known that women are more deeply involved in kin networks than men.[5] Thus, it is reasonable to expect that the experience of grandparenting differs depending on whether the link is through a son or a daughter, regardless of whether a disruption occurs. To the extent that divorce may have different implications for the two lines of descent, our survey may suggest ways in which marital disruption is changing the typical patterns of American kinship.

Remarriage may also be altering American kinship as it expands the number of kin. We will examine the stepgrandparent-stepgrandchild relationships that are increasingly common due to the large numbers of remarriages. As will be noted, however, the information in our study about these emerging relationships is less detailed than we would have liked.

Grandparents and the Process of Marital Dissolution

But first, let us examine the role of the grandparents as the process of marital dissolution unfolds. In cases in which the study child's parents had separated or divorced, we asked the grandparents to recall the events surrounding and following the dissolution. The effects of marital dissolution were quite different for the grandparents on the side of the parent who stays with the child than for the grandparents on the side of the parent who leaves the child's home. We will refer to these two groups as the "custodial grandparents" and the "noncustodial grandparents," respectively.[6] Most of the custodial grandparents were maternal grandparents, and most of the noncustodial grandparents were paternal grandparents.

The Initial Period

About seven out of ten grandparents on both the custodial and the noncustodial sides reported retrospectively that their children tried to keep their marital troubles from them before the breakup.[7] Moreover, less than half of the grandparents were told about the plans to separate before the breakup occurred. It appears, then, that adult children tend to keep their parents in the dark about their marital troubles. In our follow-up interviews, some grandparents said that they were consulted by their children before the breakup, and others said that they suspected that the marriage was not going well. But more typical was the response of a middle-class woman from a large northeastern city to the news that her daughter was separating:

> Barbara [her daughter] had kept up the facade that they were very, very happy. . . . And then one time, for about four weeks I didn't hear from her. And finally she called up and said, "John and I are separating." I hit the floor. . . . Months later, when I had a chance to talk with her, I said, "Why didn't you tell us?" She said, "I didn't want to hurt you and Daddy, because you wanted so much for me to be happy. . . ." And she said, "That's another reason, I didn't know how you'd react." And I think she always felt that I felt that she should have stayed with him rather than have the stigma of divorce. But I didn't feel that way.

The woman's daughter at first explained her silence as the result of her desire not to hurt her parents' feelings. Subsequently another reason emerged: Barbara appears to have been afraid that her parents would have exerted pressure on her to stay married. Barbara seems to have excluded her parents from the decision-making process partly in order to avoid interference from a third party with a presumed interest (which her mother denied) in maintaining the marriage. So, although Barbara initially justified her actions in benevolent terms, her silence also allowed her to remain independent of her parents while she resolved her marital crisis.

The news of an impending separation hits grandparents hard even when they know in advance that the marriage is in trouble. Grandparents may be distraught over whether to "interfere" on behalf of the grandchildren. In a follow-up interview, Mrs. Douglas, whose daughter had divorced, was read the following hypothetical story (suggested to us by Colleen Johnson):

> INTERVIEWER: *How about this situation? Mrs. Martin has been very upset for some time and doesn't know what to do. Her son's marriage is on the rocks and she knows there's a lot of fighting and unhappiness. Her granddaughter seems to be affected by it. She's more tearful and anxious. Mrs. Martin wants to avoid being an interfering grandparent, yet she wants to help her granddaughter. How can she handle that?*
>
> How can she handle it? Well, I can only tell you what I feel. What I did in the same situation. Laura was small, but ever since it happened, I always made her feel like she could come to me. And I think that's most important [breaks down crying].
>
> INTERVIEWER: *Well, it sounds like she does. It sounds like she feels she can come to you.*
>
> I feel like if you were always a good grandmother you couldn't stand back regardless . . . yet you have to be diplomatic and you have to figure it all out.

Mrs. Douglas wanted to be a good grandmother, but what does a good grandmother do in this situation? She couldn't stand back but she had to be diplomatic; she obviously wanted to intervene but felt she lacked the standing to help negotiate her daughter's marital problems. Given these constraints, the best she could do was to offer a haven for her granddaughter.

Like Mrs. Douglas, most grandparents felt they lacked the authority to intervene in the marital dispute. This is another example of the power of the norm of noninterference. Consider the response of a middle-class woman from a southern city who was asked why she did not try to prevent her daughter from leaving her husband:

> She'd been raised in a Christian home; she knew we did not believe in divorce. And he [the grandmother's ex-son-in-law] felt like if we said you've got to live with him, she would have done that. But I just couldn't be that kind of parent. Even though I do not believe in it, I could not say, hey, you've got to live with that for the rest of your life. I mean, that's somebody else's life.

Moreover, an attempt to interfere would be risky; for should it fail, it might alienate a child or a child-in-law and strain relations with the children and grandchildren after the breakup. Nevertheless, about half of the custodial grandparents and three-fourths of the noncustodial grandparents reported that they discussed with the parents the effect the breakup might have on the study child. Far fewer talked directly with the child about the breakup. The study children often were quite young at the time their parents separated, but in addition many grandparents felt that they did not have the authority to bypass the parents and talk directly to the grandchildren about difficult matters.

Noncustodial grandparents were more likely to have made an effort to prevent the divorce (46 versus 34 percent) and, as has been noted, to discuss its consequences than were the custodial grandparents. This difference might reflect the higher stakes for the noncustodial grandparents. As Vern Bengtson and Joseph Kuypers have argued, older people in general have a greater stake in maintaining generational continuity.[8] And in this instance, the noncustodial grandparents are particularly at risk of losing that continuity. They fear, with some justification as we will see, that their relationship with their grandchildren will suffer after their child moves out. Most custodial grandparents have no such fears:

> INTERVIEWER: *Did you ever think of how [the divorce] could affect your relationship with [the child]?*
>
> No, well, I knew that wouldn't because my daughter was his mother, see? I knew that wouldn't affect our relationship because she got custody of him.

When the breakup occurs, the grandparents are called upon for substantial amounts of assistance. Research on extended kin relations shows that adult children turn to their older parents for help during emergencies; and this tendency is clear in our study. Six out of ten custodial grandparents reported that they provided some financial assistance to the study child or to either of the child's parents. Moreover, three out of ten custodial grandparents said that their grandchild came to live with them (usually accompanied by the custodial parent) about the time of the breakup. Among the noncustodial grandparents, the reported levels of assistance were lower but were still substantial: four out of ten said they provided financial help, and one out of seven said that their grandchild came to live with them. Custodial grandparents were much more likely to help if they lived close by, consistent with our general finding about the importance of distance.[9] In addition, custodial grandparents were more likely to help if they reported a very close relationship with the custodial parent or if they had more education—most likely an indicator of available resources. Moreover, grandparents whose daughters retained custody of the child were more likely to help than were grandparents whose sons retained the child.

Overall, the retrospective reports of the grandparents imply that their role in the process of marital dissolution was small in the initial stages but grew when the separation occurred. Few grandparents served as advisors or confidants when marital problems first emerged; on the contrary, most were intentionally left out of the situation by their children. Once the decision to separate was announced to them, some grandparents entered the discussions about the effects of the breakup. Then, when the separation finally occurred, many grandparents provided support, especially those living nearby.

At the beginning of the process, the roles of the custodial and noncustodial grandparents were similar, though as shown previously noncustodial parents were slightly more likely to intervene on behalf of preventing a marital breakup. Soon after the separation occurred, the behavior of custodial and noncustodial grandparents began to diverge. As noted, the custodial grandparents were somewhat more likely to provide support during and just after the breakup. Moreover, the custodial grandparents were better able to maintain or increase contact with the grandchildren. Grandparents on both sides were asked whether, during the breakup, they saw the study child more often, less often, or about the same amount of time as before the marital problems began (question 92). The results are as follows:[10]

	Custodial Grandparents	Noncustodial Grandparents
More often:	43	28
Less often:	6	41
About the same:	51	31
	100%	100%

Nearly all custodial grandparents said that they saw the child either "more often" than before the marital problems began or "about the same amount of time." In contrast, noncustodial grandparents were much more likely to report that they saw the child less often than before the marital problems began. It appears, then, that even at the initial stages of the process of dissolution, many noncustodial grandparents already had experienced a reduction in the number of visits with their grandchild. Already the breakup of the marriage began to impose barriers to free contact between noncustodial grandparents and their grandchildren. Still, there was substantial variability: a number of noncustodial grandparents, as can be seen, provided assistance and saw the grandchild more often during the breakup.

The Aftermath

What are the long-term effects of a marital dissolution in the middle generation on the relationship between grandparents and grandchildren? Is the assistance provided by grandparents temporary, or do they remain an important source of aid? Does the greater role of the custodial grandparents just after the breakup translate into a greater role several years later? Do most noncustodial grandparents become peripheral, or are they able to recoup their position prior to the breakup? We can begin to answer these questions by examining the situation in 1983 of grandparents in maritally disrupted families. About four-fifths of the cases of marital disruption in the parent generation in our study occurred prior to the first interviews with the children in 1976. (The study children already were seven to eleven years old then, and most separations occur during the early years of marriage.) Thus, in most cases the 1983 interviews with grandparents took place at least seven years after the disruption.

A Special Case: Grandparents and Divorce

The grandparents were asked in 1983 whether, at the present time, they see the study child more often, less often, or about the same amount of time as before the breakup (question 93):[11]

	Custodial Grandparents	Noncustodial Grandparents
More often:	23	22
Less often:	37	58
About the same:	40	20
	100%	100%

Among the custodial grandparents, the responses were mixed, with substantial proportions in all three response categories. The majority had at least maintained the level of contact established before the breakup. The most noticeable difference from the question about contact during the breakup is that many more custodial grandparents—although still a minority—said they saw the study child less often than before the breakup. Further analyses showed that custodial grandparents who reported less contact now or about the same amount now tended to live farther from the study child than those who reported more contact now. Thus, the apparent decline in contact for a minority of custodial grandparents could have resulted from some families moving farther away from the grandparents after the dissolution. Additionally, however, it must be remembered that the study children had become teenagers, who may have had less contact with kin than they did when they were younger.

As for the noncustodial grandparents, it can be seen that the lower levels of contact that were present during the breakup persisted for many of them. They still were more likely to report a decline in contact than were the custodial grandparents. Nevertheless, a distinct minority of the noncustodial grandparents reported "more contact" than before the breakup, reflecting the variable situation of this group. Moreover, the difference between custodial and noncustodial grandparents is smaller than it was during the breakup. Apparently, those noncustodial grandparents whose relationships survived the breakup managed to retain ties over time.

Further tabulations showed that in 1983 the custodial grandparents still were more likely to be living with a study child (12 versus 4 percent) or living within one mile (10 versus 2 percent) than were the noncustodial grandparents. Moreover, among all grandparents who were not living with the study child, noncustodial grandparents tended to see their grandchildren much less often than custodial grandparents. Consider these responses to a question about how often they had seen the child in the past twelve months:[12]

	Custodial Grandparents	Noncustodial Grandparents
Almost every day	10	2
Two or three times a week	13	0
About once a week	16	14
Once or twice a month	17	22
Once every two or three months	13	14
Less often	31	48
	100%	100%

At one extreme, 23 percent of the custodial grandparents had seen the study children at least two or three times a week, compared to just 2 percent of the noncustodial grandparents. At the other extreme, 48 percent of the noncustodial grandparents saw the study children less often than once every two or three months, compared to 31 percent of the custodial grandparents. In addition, 33 percent of the custodial grandparents still were providing financial support to the child's parents, compared to only 10 percent of the noncustodial grandparents.

This, then, was the typical situation several years after the breakup: the noncustodial parents tended to live farther away from their grandchildren than did custodial grandparents, to see them less often, and to provide less financial support. Moreover, we can safely assume that our survey findings underestimate the true reduction in contact among noncustodial grandparents. Recall that we were given the names of the thirty-two noncustodial grandparents we interviewed by the parent who was residing with the child in 1981. A close examination of these thirty-two families revealed that the noncustodial parents (usually the fathers) were more likely to be seeing the study children regularly than would be expected from national figures.[13] Therefore, the noncustodial grandparents we were able to interview tended to be the fortunate ones whose sons still were in touch with the grandchildren after the divorce. The less fortunate ones, whose sons did not have much to do with their children, are underrepresented because we were not given their names and addresses, as often by the ex-daughters-in-law.

Why, if some of their children were maintaining contact, did the noncustodial grandparents we interviewed still see less of their grandchildren, on average, than other grandparents? The major reason was distance: they lived farther away from their grandchildren, probably because the custodial parent had moved away. Among the four out of ten who said they lived more than one hundred miles away, all reported that they saw the grandchild "less often" than "once every two or three months"—the lowest response category offered—despite sometimes substantial contact between the grandchild and the noncustodial parent. In addition, some noncustodial grandparents perceived resistance from the custodial parent to more frequent visits. One out of eight responded affirmatively to the question, "Do you ever find that you don't spend as much time with [the child] as you would like because . . . either of [the child's] parents don't want you to see [him/her] more?"

Those noncustodial grandparents who had a continuing link through their children and who lived nearby seemed to have had the best chance of retaining regular contact with their grandchildren.[14] Many noncustodial grandparents, we suspect, either lack a continuing link through their children or live relatively far away—or both. They are at risk of losing their regular ties to the "kin-keepers"—the daughters-in-law in the middle generation who are central to keeping in touch.[15] To be sure, a minority still maintains regular contact with the grandchildren, and a small number played an enhanced role in their grandchildren's lives. But we would speculate that many, perhaps most, noncustodial grandparents have what we might call ritual contact with their grandchildren: they see their grandchildren infrequently, on special occasions when the child happens to be visiting his father or when they are able to make special arrangements for a visit with the custodial parent.

Remarriage

Most divorced spouses remarry, although the rate of remarriage has been declining recently.[16] Does the remarriage of the custodial parent alter the role of the custodial grandparents? Our data show that the percentage of custodial grandparents who were living with the study child was sharply lower (3 percent) in cases where the parent had remarried, compared to cases in which the parent had not remarried (17 percent). When parents remarry, they tend to form independent households with their new spouses, if possible. In addition, the percentage of custodial grandparents providing financial support was lower (22 percent) in cases where the parent had remarried than when she had not remarried (41 percent). When divorced mothers remarry, their economic situation typically improves because of the addition of a (usually higher) male income to the household; moreover, divorced mothers who are not receiving financial support may have a stronger motivation to remarry. Yet the frequency of contact between custodial grandparents and their grandchildren was not affected much by a remarriage.[17] In sum, remarriage reduced the support provided by custodial grandparents, consistent with the greater economic resources available to families of remarriage; but levels of contact did not change markedly after a remarriage.

Unfortunately, the number of cases of noncustodial grandparents was too small to support a similarly detailed analysis of the survey data. We explored the effects of remarriage on noncustodial grandparents in our follow-up interviews. The results tentatively suggested that the remarriage of the custodial parent further complicates the situation of the noncustodial grandparent. One woman, whose granddaughter, Susan, lived with her ex-son-in-law (Susan's biological father) and his second wife (Susan's stepmother), told an interviewer:

> It's hard [Susan not living with her biological mother], it's really hard, and I realize that; and so we, my husband and I, we just say we don't want to make it harder. Because we don't want to cause any big ripples, you know, in her relationship with her stepmother. Like for us to insist that she come visit if it doesn't suit the stepmother. . . . It would just cause problems.
>
> INTERVIEWER: *So you. . . .*
>
> I don't see her as much as I would like to see her.
>
> INTERVIEWER: *Because of possible conflict with the stepmother?*
>
> Yes, that's right. I just don't want it and so I just, rather than have Susan . . . feel like she's in the middle, I don't like that, I don't want her to feel that way.

Thus, in addition to overcoming distance (she lives more than one hundred miles from Susan) and a daughter who rarely sees Susan, this grandmother must now contend with a stepparent who may be trying to unify her new stepfamily around her kin and her husband's kin. As a result, she feels more removed from Susan than ever.

Divorce and Kinship

Having shown how the process of marital dissolution affects custodial and noncustodial grandparents, we are now ready to consider whether divorce is altering the American kinship system. In this section, we will examine whether the ties between grandparents and grandchildren are different when adult children have divorced than when they are in a first marriage. Since intergenerational relations may differ, regardless of divorce, according to whether the link is through a son or a daughter, we will look separately at the maternal and paternal sides. This division yields four groups: maternal and paternal grandparents whose children are in intact first marriages, and maternal and paternal grandparents whose children have divorced. We compared these four groups using five indicators of the grandparent-grandchild relationship: geographical distance, frequency of contact, financial support, the scale of parentlike behavior, and the scale of exchange of services. (The two scales have been discussed extensively in previous chapters.) The results of these comparisons are presented in table A–7 in appendix 3.

We looked first at grandparents in nondisrupted families only, and we found few differences on any of the indicators between the maternal and paternal sides.[18] In the absence of a marital disruption, it would appear that the level of involvement of the two sets of grandparents is, on average, similar. The hypothesis that grandparenting differs according to whether the link is through a son or a daughter is not borne out. Although grandmothers still may be more active in kin networks than grandfathers (as, indeed, other analyses of this survey suggest), we find no evidence that grandparents are more involved with maternal grandchildren than with paternal grandchildren. Instead, as we argued in chapter 5, the underlying principle of intergenerational relations in nondisrupted families would seem to be equity: in general, the maternal and paternal grandparents are given equal access to the grandchildren and provide similar amounts of assistance. This pattern is consistent with the bilateral nature of American kinship, which gives equal emphasis to both bloodlines.

(But as noted previously, our data may understate differences between maternal and paternal grandparents. Fortunately, we can check our interpretations by tabulating responses from the 724 currently married, never-disrupted mothers who were interviewed in the 1981 parent survey. Sixty-one percent reported that their parents lived within an hour's drive, and 62 percent reported that their husbands' parents lived within an hour's drive—nearly identical percentages. Concerning assistance, 19 percent reported that they had received "help such as childcare, errands, housework, or home repairs" from their parents in the past few weeks, and 15 percent reported receiving such help from their husbands' parents. This modest difference also was evident in the responses to a question about the receipt of "financial help" during the past year: 12 percent received financial help from their parents; 8 percent received it from their husbands' parents. Only on a question concerning receipt of "moral support such as advice or encouragement" during the past few weeks was there a substantial difference, with 51 percent receiving moral support from their parents compared to 32 percent receiving moral support from their husbands' parents. Overall, these responses are consistent with our conclusion that, in the absence of marital disruption, the maternal and paternal sides have similar access to the grandchildren and provide similar levels of assistance; though they also suggest that women may be emotionally closer to their mothers than to their mothers-in-law.)

When comparisons are made between grandparents in disrupted versus nondisrupted families, on the other hand, sharp differences are found. Among all maternal grandparents, those in disrupted families were more likely to be living with the study child or to be seeing that child almost every day, were more likely to be exchanging services, were more likely to be providing financial support to the child's parents, and were much more likely to be engaging in parentlike behavior.[19] The differences in exchange and parentlike behavior suggest that the quality, not just the quantity, of grandparent-grandchild relations is different after a disruption. We view the high levels in the disrupted group of behavior usually reserved for parents as an especially sensitive indicator of the quality of the relationship. Almost universally, as we have mentioned in previous chapters, the grandparents we spoke to endorsed the norm of noninterference—the idea that grandparents ought not to interfere in the ways that their children are raising the grandchildren. Consequently, those grandparents who were disciplining, advising, and correcting their grandchildren with regularity had crossed a sensitive boundary in intergenerational relations.

The results of our survey suggest, then, that the relationship between maternal grandparents and their grandchildren is deeper, in many cases, in families that have experienced a marital disruption. Many of the maternal grandparents in the disrupted group lived with their grandchildren for an extended period of time; those grandparents who were not residing with their grandchildren saw them about as often as maternal grandparents in non-maritally-disrupted families. Long after the breakup, the grandparents in the disrupted group still were providing relatively frequent financial support. Moreover, the quality of the relationship was often different: grandparents in the disrupted group did more things for their grandchildren than other maternal grandparents and received more in return. And many more of them had the authority to act toward their grandchildren in ways normally restricted to parents.

Among paternal grandparents, the effects of marital dissolution were quite different. Those who had experienced a dissolution tended to live farther from the study children and to see them less often.[20] Thus, disruption appears to have reduced the quantity of interaction. But unlike the maternal side, the quality of the interaction seems to have been relatively unaffected: both paternal groups were similar in the extent of parentlike behavior and the amount of financial support provided (for both indicators, far less than in the maternal-disrupted group), although the disrupted group was less likely to be exchanging services. Of course, as we have noted previously, the modest sample of paternal grandparents from divorced families probably overrepresents grandparents who had a continuing relationship with their ex-daughters-in-law after the disruption. Even so, there is no evidence of deeper intergenerational ties on the paternal side of families that have experienced a disruption. On the contrary, the picture is one of declining frequency of contact and a quality of interaction that is lessened somewhat or, at best, maintained.

These comparisons show that marital disruption tends to deepen the quality of relationships between maternal grandparents and their grandchildren. This conclusion is based largely on the finding of a higher level of parentlike behavior among grandparents in the maternal-disrupted group. But can we be sure that this higher level of parentlike behavior is not just a reflection of other characteristics of this group—its racial composition, perhaps, or its income level? In order to answer this question, we analyzed the determinants of parentlike behavior among all maternal grandparents, using a method—multiple regression—that allowed us to control for the effects of several potential causes simultaneously. This analysis was quite similar to our analysis in chapter 5 of the determinants of the involved style of grandparenting—only this time the data were restricted to maternal grandparents. The results demonstrated once again that even after statistical controls for such important predictors of parentlike behavior as frequency of contact, the age of the grandparent, and race, maternal grandparents still were more likely to engage in parentlike behavior with their grandchildren if their daughters were divorced or separated. The difference in adjusted mean scores between the disrupted and nondisrupted groups was nearly one point on this five-point scale.[21] Divorce and separation do appear to make a difference in the roles grandparents play.

The obvious explanation for the different effects of marital disruption on the maternal and paternal sides is the prevailing pattern of child custody arrangements. Indeed, custody is so confounded with lineage in our study that we cannot separate the two. Following the national pattern, nearly all of the maternal grandparents in disrupted families in our study are on the custodial side; and most of the paternal grandparents in disrupted families are on the noncustodial side. On balance, then, being on the maternal, custodial side after a disruption would appear to enhance grandparent-grandchild relations, but being on the paternal, noncustodial side would seem to diminish them.

Because we believe that our sample of paternal, noncustodial grandparents overrepresents those with a continuing link to their former daughters-in-law, we strongly suspect that the true picture for paternal grandparents is even bleaker than our data suggest. The breakup of a son's marriage, it seems, often results in a diminution of the relationship between paternal grandparents and their grandchildren—a loss that is painful for the grandparents involved. Still, it is important to note that most paternal grandparents continue to see their grandchildren after a marital disruption, albeit less often. The ritual ties between the generations appear to survive the divorce process, although most paternal grandparents seem to be removed from the day-to-day lives of their grandchildren after a divorce.[22] Even this modest degree of contact may be seen as a considerable amount, given that many of the noncustodial fathers have very limited ties to their children.

Most of the public discussion of this problem has centered on grandparents who are denied access to their grandchildren by the custodial parent and who must resort to legal intervention to obtain visitation privileges.[23] Although our sampling procedure tended to screen out these grandparents, we did find evidence that parents discouraged frequent contact in a minority of cases. Yet we would caution that noncustodial grandparents more often face other serious difficulties. Foremost among these is distance. Many custodial parents move after the breakup, and these moves can take the grandchildren beyond easy visiting distance. Moreover, the position of the noncustodial grandparent may become even more problematic when the custodial parent remarries, thereby adding a new set of kin to the grandchildren's family. Finally, the grandparents' son—the noncustodial parent—may have dropped out of his children's lives, making it awkward and difficult for the grandparents to arrange occasions for visits. Thus, even when the custodial parent would allow the noncustodial grandparent reasonably regular access to the grandchildren (which we suspect is true in most cases), the noncustodial grandparent may face other difficulties. These additional difficulties are not the type that are suited to legal intervention. If efforts to help noncustodial grandparents see more of their grandchildren are narrowly focused on grandparents' legal right to visitation privileges, many noncustodial grandparents will not be helped at all.

Stepgrandparents and the New Extended Family

A divorce in the middle generation often brings further changes to the family lives of grandparents because it is likely that their divorced children will remarry.[24] And if an adult child marries someone with children from a previous marriage, the grandparent suddenly will become a stepgrandparent. The meaning of this new kinship tie varies greatly according to the particular circumstances of the new extended family. Consider, for example, two hypothetical grandmothers. The first has a son who shares a household with his second wife and her three-year-old son from a previous marriage. The second has a daughter who marries a man with a teenaged son who still lives with his mother in a neighboring state. The first grandmother probably would have a much closer and more kinlike relationship with her stepgrandchild than would the second one.

Unfortunately, we were not successful in obtaining the names of the stepgrandparents of the study children—we completed only twelve such interviews. The very reluctance of remarried parents to provide us with the names of stepgrandparents may be a sign that stepgrandparents often are not well integrated into family life after remarriage. Nevertheless, one-third of all the grandparents interviewed in the survey had at least one stepgrandchild—a striking example of how many families are touched by divorce and remarriage. We included questions in the survey about the difference between step- and biological relations with grandchildren (numbers 97 through 99g). But we quickly found that these few questions provided very limited insight into complex step-relationships. In our follow-up interviews, therefore, we probed in more detail about what it means to be a stepgrandparent.

Two key factors emerged: (1) the age of the stepgrandchild at the time he or she first became a member of the grandparent's extended family, and (2) whether or not the stepgrandchild lived full-time with the grandparent's adult child. At one extreme, consider Mrs. Stevens, who described in chapter 4 her relationships with her two twenty-four-year-old granddaughters. One of these granddaughters is the child of Mrs. Stevens's son. The other is a stepgrandchild, who came into Mrs. Stevens's family when her daughter married a man with a two-year-old child. Mrs. Stevens watched both girls grow up. Asked what the difference was between having a young stepgranddaughter and a young granddaughter, she replied:

> Really none, because I love her [the stepgranddaughter] just as much. I mean it was great because I had seen two little girls grow up, both almost the same age. . . .
>
> INTERVIEWER: *So they are almost like twins?*
>
> Yeah, I dressed them alike when they were young. Everything I bought for them, I bought alike. It was funny—when the one got older, when she was around me, my own granddaughter was around me much more. She said, "Oh, please, Grandma, don't buy the same things"—you know.
>
> INTERVIEWER: *But to you they are the same?*
>
> I didn't feel any different. . . . They were cute together. There's a picture of them on the wall there. The blond one is my own, but I feel like the other one is just as much my granddaughter. Actually, I don't even like to say, you know, I don't know what you call it, stepgranddaughter or what really.

Despite her distaste for the term "stepgranddaughter," Mrs. Stevens matter-of-factly referred to her biological granddaughter as "my own" or "my real granddaughter" during the interview. She was aware of the genealogical difference between the two granddaughters, but it seemed to carry little emotional or practical significance. The more important difference was that her daughter later moved to another state so that her biological granddaughter "was around me much more." Perhaps, deep inside, Mrs. Stevens does feel closer to her biological granddaughter, but in her actions and her expressions of sentiment she showed no favoritism. Her stepgranddaughter had arrived on the scene at such a young age that Mrs. Stevens had been able to play the role of grandmother during the enjoyable preadolescent years—the fat part of grandparenting. Moreover, Mrs. Stevens faced little competition from her stepgranddaughter's maternal grandmother: her stepgranddaughter saw this other grandmother only on holidays and special occasions.

In contrast, Mrs. Scott, a woman from a midwestern city,

acquired two stepgrandchildren when they were in their early teens. Her daughter had married their father, a widower. The two stepgrandchildren, however, lived not with their father but with his parents. Mrs. Scott explained:

> That had nothing to do with the marriage, because he just couldn't handle it alone after his wife died. So his mother and father took the children and finished raising them.
> INTERVIEWER: *And what do they call you?*
> Harriet. I insisted on that. They started by calling me Mrs. Scott. . . . But from the beginning, you realize, these children were in their teens and it was hard to accept somebody from an entirely different family and they didn't know me from Adam. . . . Now if they were smaller—you know, younger—it would have made a difference. But I think they think of me as a loving stepgrandmother. Now this I'm sure of. . . . And that's the way I like it, you know, because [her daughter] didn't try to take the place of their mother, I didn't try to take the place of their grandmothers, you know. I wouldn't think of doing that. You can't take another person's place. Just make your own little passage into their heart with love, and then you can try for a little piece of it that's all your own; you don't have to take it away from any other grandmother.

If they were younger it would have made a difference because they would not already have formed a close attachment to their biological maternal grandparents. And they would not have known Mrs. Scott initially as the mother of someone their father was dating:

> I had seen them before Diana and George married, so naturally they called me Mrs. Scott. . . . So then Diana and George got serious with each other and were married, and it's hard after calling somebody "Mrs." for a long time to switch to something else.

Indeed, the transition from acquaintance to stepgrandparent can be difficult when teenaged grandchildren, whose sense of family and place is established, are involved. Mrs. Scott wisely backed off from competing with the existing grandparents and instead sought her niche as a loved and loving stepgrandparent. She seems to have succeeded.

If our follow-up interviews are an accurate guide, the rule would seem to be that the older the children are when the remarriage occurs, the less likely are the stepgrandparents to play a role similar to that of biological grandparents. Mrs. James, who lives in a large northeastern city, has a daughter married for the second time to a widower with three children: one in the first year of college, one in the last year of high school, and one in junior high school.

> INTERVIEWER: *What kind of a relationship did you establish with his children?*
>
> I didn't see them that often. We got along.
>
> INTERVIEWER: *Do you think they think of you as a real grandmother?*
>
> Well, I don't know. I don't think they do. See, their grandmother, my son-in-law's mother, lives there close by. Of course they are nice. Of course they don't consider me a real grandmother.
>
> INTERVIEWER: *How do you feel about them? Do you feel any difference between them and your natural grandchildren?*
>
> Oh yes, I'm much closer to my own grandchildren.

Mrs. James has not established much of a relationship with her stepgrandchildren, nor does she think it possible to do so, given their independent lives and their close ties to biological grandparents: "Of course they don't consider me a real grandmother."

We have noted at several points in this book how the relationship between grandparents and grandchildren mirrors the parent-child relationship. Similarly, the relationship between stepgrandparents and stepgrandchildren mirrors the stepparent-stepchild relationship. Stepparents also can have difficulty defining their role, particularly when the stepchildren they acquire are older or live elsewhere or have close ties to both biological parents.[25] That stepgrandparents—who have less control over the children and do not live with them—have similar problems is understandable.

Moreover, the strong influence of the age of the stepgrandchild at the time of remarriage leads us to suspect that more basic processes of emotional attachment may be at work. Research has shown that young children establish strong bonds of attachment to their parents that have great importance for their emotional well-being. Young children rely on their parents for security and they protest when that security is threatened, as when an infant cries when his mother leaves the room. The child's attachment becomes somewhat reduced after the preschool years and appears to decline steadily thereafter.[26] Perhaps, and this is speculation on our part, young children can form similar, though maybe less intense, bonds of attachment to grandparents in a way that older children cannot. If so, then it may be that a child who enters a family after a certain age—perhaps as early as four or five—cannot establish the kind of attachment that leads both the child and the stepgrandparent to feel that their relationship is "real" or "natural."

This is not to suggest, however, that the relationships between stepgrandparents and stepgrandchildren are necessarily problematic. On the contrary, the process of adjustment appears to proceed smoothly in most cases. Most of the remarried persons studied by Furstenberg and Spanier in central Pennsylvania reported that the process of introducing their children to their partners' families was comfortable and pleasant; only one out of ten described serious complications.[27] In our survey, similarly, only 9 percent of the stepgrandparents agreed with the statement "You've had problems accepting your stepgrandchildren as grandchildren."[28]

Yet the survey data also showed the variation in step-relationships that was apparent in the follow-up interviews. About one-third of the stepgrandparents agreed with such statements as "Your stepgrandchildren can't think of you as a real grandparent," "It is harder for you to be a stepgrandparent than a grandparent," "You feel differently about your natural grandchildren than your stepgrandchildren," and "It is generally harder for you to love your stepgrandchildren than your own grandchildren."[29] And 56 percent agreed with the statement "You tend to see less of your stepgrandchildren." Moreover, a difference existed in forms of address: 28 percent of the stepgrandparents reported that their stepgrandchildren called them by their first names, by nicknames, or by formal names (such as "Mrs. Scott") rather than by kinship terms (such as "grandma" or "mom-mom" or a foreign-language version of these). In contrast, only 8 percent of their natural grandchildren did not use kinship terms.

We strongly suspect that grandparents whose stepgrandchildren were older when they entered the family, or whose stepgrandchildren did not live with the grandparents' adult child, were more likely to agree with the statements about problems and were more likely to be addressed by non-kinship terms. The follow-up interviews unambiguously suggested as much. But we cannot test our belief with the survey data because, having underestimated the complexity of step-relationships, we did not ask how old the stepgrandchildren were when they entered the family or where they lived. (Recall that most of the interview was focused on the study child and that most of the stepgrandchildren referred to in these questions were not the study children. It seems that whenever survey researchers finish a study, they belatedly discover a question or two they cannot believe they neglected to ask. These are ours.)

Clearly, there is more to learn about the new intergenerational relationships that have formed in so many families because of remarriage. Remarriage can result in an expansion of kin: Furstenberg and Spanier, for instance, found that children who had more contact with stepgrandparents had no less contact with their biological grandparents.[30] What occurred instead was an increase of the number of grandparents they saw from two to three sets. There is the possibility of a fourth set, if the noncustodial parent remarries and the children visit his new wife's parents. And for those few children whose custodial parent marries three times (or lives with a partner after two divorces) there could even be a fifth set. It is hard to imagine, however, that all of these intergenerational relationships are deep and meaningful. Certainly our follow-up interviews show a great variation in the quality of relationships. We would venture to guess that very few stepgrandparents have a high degree of parentlike authority over their stepgrandchildren and that the exchange of services is typically quite modest. The exceptions, we suspect, would occur in families where the stepgrandchild is structurally similar to an adopted child—one who enters the family in the first few years of life and lives with them full time. Otherwise, most relationships probably range from the companionate to the remote. Without doubt, remarriage is distributing a smaller pool of children among a larger group of grandparents. But the nature of these new, postdivorce relationships is still an open question.

A Summing Up

The data and the case studies presented in this chapter allow us to draw some lessons about the intergenerational aspects of American kinship in an era of frequent divorce. We would suggest that the principle governing the restructuring of kinship ties after divorce and remarriage is the same as the more general one that determines the nature of kinship ties throughout American society: a great deal of discretion is given to individuals to distinguish between relatives and nonrelatives. Blood and marriage (and now perhaps cohabitation) circumscribe the available pool of kin, but within this pool it is up to individuals to cultivate and maintain ties.[31] Thus, whether or not a grandparent has a continuing tie to a grandchild after a divorce—or whether or not a stepgrandparent establishes a strong tie to a stepgrandchild after a remarriage—depends on the actions of both parties. Being a grandparent after divorce and remarriage is in part an achieved status.[32] After a divorce, some of the custodial grandparents in our study acted more like parents and exchanged more goods and services with their grandchildren than before. But others—perhaps those who lived far away or did not get along with the grandchild's parents or were not needed or were not able to help—were unable to maintain their relationships with their grandchildren after the breakup, despite their biological claim of kinship. On the other hand, although most noncustodial grandparents seemed to have a diminished relationship with their grandchildren, a few managed to retain close ties through their own actions. Stepgrandparents showed a great diversity in the nature of their ties to their stepgrandchildren. There are no fixed rules, no hard and fast outcomes. A genealogical relationship is no guarantee of a sociological one: "Blood doesn't make a family," as one stepgrandparent said. Instead, what makes a family and what cements intergenerational ties is the work of maintaining relationships. Another grandmother commented:

> Kin to me means when you act like kin. You have a closeness, you have a feeling for the person. When you say, "This is my cousin," your cousin means something—that role, you know.

In addition, the findings in this chapter suggest that intergenerational ties are often latent in the kinship system. When the family is functioning well, grandparents usually are content to take a quite secondary role to the parents. They restrict themselves to pleasant interaction and to limited exchanges of services. They may be regular visitors and important symbols of family continuity, but their day-to-day contribution is modest. Still, they watch for signs of trouble, and they often offer substantial assistance when a crisis such as divorce occurs. They constitute a latent source of support for the family system that often becomes manifest after a divorce in the middle generation.

Yet despite the diversity of responses to divorce and remarriage, some central tendencies can be observed. For grandparents, the most important tendency is as follows: if your daughter's marriage breaks up, your relationship with your grandchildren probably will be maintained or even enhanced; but if your son's marriage breaks up, your relationship with your grandchildren is likely to be diminished in quantity and quite possibly in quality as well. This tendency is dependent in large part upon existing patterns of child custody, in which most children remain in the custody of their mothers. If current custody patterns continue, divorce may result in a matrilineal tilt in intergenerational continuity, as other researchers have suggested.[33] Should joint custody become more common in the future, however, paternal and maternal intergenerational ties probably would become more equal.

For grandchildren, divorce and remarriage can bring about the potential for expanding the number of relationships with grandparents. But the quality of the new step-relationships is highly uneven. The most important central tendency is that after divorce, children may have a deeper relationship with their maternal grandparents than they had with either set of grandparents prior to the breakup of their parents' marriage. Divorce has become a common, severe family crisis. But far from uniformly destroying the bonds of kinship, divorce appears to strengthen intergenerational ties along the maternal line. As a result, children of divorced parents may have stronger ties to some of their grandparents than children from nondisrupted marriages have to any of their grandparents.

7

The Influence of Grandparents on Grandchildren

One of the grandparents we spoke with told us the following story:

> We were on a television program, and the topic was grandchildren don't need grandparents—they don't need our wisdom. Through technology and going to colleges and learning so much, that the need to come to the parent and the grandparent for advice was becoming extinct. My opinion is, though, the way we live today, the human factors are the same as they were years ago. And you can't bypass us and what we have to offer by learning technical skills or getting Ph.D.'s.
>
> INTERVIEWER: *So you disagree—you think you do have something to offer.*
>
> Yes, we have an awful lot to offer. . . . What we have to pass down to each generation is a sense of values; and if we

don't have values we're going to butt our heads against the wall of destiny.

Few would disagree. We wondered, however, to what extent grandparents do pass down values to their grandchildren. More generally, how much influence do grandparents have over their grandchildren's attitudes and behaviors? In previous chapters we have looked at the nature and significance of intergenerational ties for grandparents. In this chapter we will focus on the opposite end of the life cycle, asking how much difference grandparents make to grandchildren.

Past research is of little help. There are only a handful of studies on the effects of grandparents on children, and most of these focus on what grandparents are supposed to do, or have the potential to do, rather than on what they actually do. In previous chapters of this book, we have tried to fill in the gaps in our knowledge about the kinds of support grandparents provide; but there is still virtually no information available about how this support affects the development of children. The few sociological studies we could find tended to concentrate on immigrant and minority families. For example, in his classic book about black families, E. Franklin Frazier described the prominent role of black grandmothers in childrearing in a chapter entitled "Granny: The Guardian of the Generations."[1] Contemporary students of black family life have shown that this pattern of caretaking survives today in the black community—as, indeed, the findings discussed in chapter 5 corroborate.[2] Yet even this body of research makes only incidental reference to how black children's behavior is affected by their relationships with their grandparents.

Nor is the large body of literature on child development of much assistance. Developmental psychologists, though centrally concerned with the process of childrearing, have largely ignored the involvement of extended kin. The few studies that exist are again mainly confined to minority families. This void is illustrated by a recent review article on grandparents by psychologists Barbara Tinsley and Ross Parke. In an admirable attempt to convince their colleagues to pay attention to the influence of grandparents on children, they cite well over one hundred articles about family and kinship, almost none of which, unfortunately, has any information about what difference grandparents actually make. As a result, they are left to conclude only that grandparents "have the potential for significantly influencing the developing child," without demonstrating whether that potential is ever translated into reality.[3]

The findings presented in previous chapters demonstrate, in fact, that most grandparents do not play a major role in the rearing of their grandchildren. Yet it may still be true that grandparents hand down important values, much as the grandparent quoted at the beginning of this chapter urged. We were able to find a small number of studies that examined whether grandparents influence the attitudes of their children and grandchildren. As we will discuss later in this chapter, most studies focus on the issue of whether values are transmitted directly from grandparents to grandchildren, on the one hand, or whether they are transmitted from grandparents to their own children, who in turn pass them on to the grandchildren.

All in all, this is not an impressive body of evidence—perhaps an indication that, in practice, grandparents do not have much influence on their grandchildren's attitudes and behavior. In order to address this issue with our survey data, we must see if children respond differently when grandparents are actively involved in their upbringing than when they are not. Because of the design of our study, there are limitations to the comparisons we can make. Ideally, we would have interviewed grandparents and grandchildren at the same time, or perhaps interviewed the grandparents prior to the interview with the children. Unfortunately, when the National Study of Children was planned in the mid-1970s, the possibility of extending the study to grandparents had not yet been raised. Consequently, when the children and their parents were interviewed in 1976 and again in 1981, little information was collected about the relations with their grandparents. And, of course, our 1983 interviews with the grandparents followed the 1981 children's interviews by nearly two years. We are forced to rely exclusively on the grandparents' reports to gauge the level of contact and involvement with their grandchildren. There is a small amount of information collected from the 1981 children's survey about the importance they accord to their relationship with grandparents, but even this information is not necessarily specific to the particular grandparents we interviewed. Therefore, we will be splicing together strands of information from the 1981 child and parent interviews and the 1983 grandparent interviews to get a general picture of the significance of grandparents in the socialization process. Still, given the almost total lack of information on the topic, the modest contribution our study can make should be of interest.

We will examine the extent of grandparental influence from several vantage points, beginning with a look at some items in the children's survey that provide an indication of the importance of grandparents. We will then study the degree of similarity between the values of grandparents and their grandchildren. Finally, we will analyze evidence on whether grandparents influence the social and psychological adjustment of their grandchildren.[4]

The Importance of Grandparents as Reported by Grandchildren

Although the adolescents in the National Survey of Children were never asked directly about their relations with their grandparents, three items in the 1981 survey provide valuable information. First, they were asked whom they would most likely turn to for help in the event of a personal problem. The question was asked in an "open-ended" manner; that is, no response categories were provided, but rather the interviewer wrote down whatever the child said. The typical response was that they relied on either a parent or a sibling. Just 2 percent of the children mentioned that they would seek help from a grandparent. Of course, this does not mean that grandparents are never consulted when help is needed, but it does show that grandparents are only rarely a primary source of psychological support for the child. This result is consistent with the in-depth study of three-generation families carried out by Reuben Hill and his colleagues. As noted in chapter 4, they found that grandparents are not high on a list of persons whom young people consult when a personal problem arises.[5]

Second, elsewhere in the interview, the children were asked if there was anyone else living outside the household whom they consider to be a parent or stepparent. All children were asked this question, regardless of whether they were living with two parents. Seven percent of the adolescents listed a grandparent as a surrogate or supplementary parent. This figure is undoubtedly an underestimate of the proportion of children who regard a grandparent as a parent figure. Since only the first person mentioned by the child was recorded, we suspect that grandparents were often not listed because many of the children who came from disrupted families first mentioned their parents living outside the home.[6]

Third, perhaps the best indication of the significance of grandparents came from an open-ended question that asked the children to say whom they included in their family. In this question, children were permitted to list as many relatives as they liked, although the interviewers were instructed not to probe for additional names but to restrict the list to names that were spontaneously mentioned. Fifteen percent of the children included at least one grandparent.

Predictably, children who included a grandparent in their family listing were much more likely to regard a grandparent as a parent surrogate (16 percent versus 5 percent) and also more inclined to look to a grandparent for help with a personal problem (8 percent versus 1 percent). Consequently, it seemed reasonable to combine the three items together into an index that measures the importance of grandparents for the children in our study. Of course, we cannot determine from the data whether the grandparent or grandparents to whom the child was referring were the same grandparents whom we interviewed, but for reasons that will become apparent shortly, it seems plausible that the overlap is great. Overall, 20 percent responded positively to at least one of the three items asking whether they regarded a grandparent as important in their lives, and 3 percent gave at least two positive responses.

Who were the children who assigned importance to grandparents? We examined a large number of social and demographic characteristics that might predict whether grandparents would be salient to the child.[7] Consistent with our previous results from the grandparent survey, we discovered that income and religiosity did not strongly affect the child's score on the index measuring the importance of grandparents to the child. The children of better-educated mothers, however, had higher scores: even after statistical controls for such factors as marital disruption, proximity, and assistance, an estimated 23 percent of children whose mothers had attended college answered positively to one or more items, compared to 10 percent of children whose mothers completed eight or fewer grades of school.

Surprisingly, black children were not more likely to regard a grandparent as a significant figure. But we believe that this index, for reasons previously mentioned, tended to underestimate the salience of grandparents in disrupted families; and black children more often lived in such families. Moreover, black grandparents who were living in the household were excluded from being listed, as the question only pertained to nonresidential surrogate parents.

As we discovered earlier, the proximity of grandparents to

the child is a powerful predictor of the nature of the relationship. In the 1981 parent interview, adults were asked if they resided within an hour of their parents, their in-laws, and, if they were previously married, their former in-laws. Twenty-five percent of the children whose parents reported that they lived within an hour of one or more sets of grandparents answered at least one of the index questions positively, compared to 11 percent of those whose grandparents all lived an hour or more away.

In addition to the information on proximity, the parents were also asked how much help they received from each of the different sets of grandparents (that is, their parents, their in-laws, or their former in-laws). They were queried separately about the amount of financial aid, moral support, and practical assistance provided by the senior generation. Financial aid from grandparents did not increase their standing in the eyes of their grandchildren (perhaps because the assistance was not known to the adolescents), but other forms of aid were related to our index of whether children saw their grandparents as important figures. Combining the information on proximity and assistance, we found that among children who had two or more sets of grandparents living close by who provided financial, practical, or moral support, 32 percent answered at least one of the index questions positively. In contrast, just 9 percent mentioned a grandparent as a significant figure when none of the grandparents lived close by or played an important helping role in the family.

This information, which is based on data from the parent and child interviews, supports our earlier findings derived from the grandparent interviews. The picture looks very similar regardless of who answers the questions: grandparents play an important role in the child's life when they live close by and have a functional role in the family. When they do, a very sizable proportion of children include grandparents as family members or regard them as parentlike figures.

Yet in the data presented so far, we cannot distinguish which grandparents are being referred to by the children or, specifically, whether they happen to be the grandparents we interviewed. It turns out that when we confine our analysis to the 510 grandparents in the 1983 interviews and their related grandchildren, the results are fairly similar to those from the 1981 parent and child surveys alone. Grandparents in our 1983 survey who reported living close by or who achieved a high score on our measures of parentlike behavior and exchange had grandchildren in the 1981 survey who were more likely to assign importance to grandparents in their family lives. Of course, we still cannot be certain that the children are exclusively referring to the grandparents in our study, but the association strongly suggests that grandparents achieve a more prominent place in the child's life when they are actively involved in the day-to-day life of the family. Otherwise, they are seen as rather marginal figures.

The Intergenerational Transmission of Values

When grandparents manage to play an active role in the child's life, do they have any measurable effect on the development of their grandchildren? Can we detect the grandparent's influence on the child's attitudes or behavior? Obviously, our study was not designed to explore this question very fully, but we can get some indication of the role of grandparents in socializing the child by investigating their influence on the children's values. We have already seen that very few children actively seek their grandparents' advice when they have a personal problem. But this does not mean they do not look to the senior generation for moral guidance or that the family elders do not have any influence on their opinions. As mentioned earlier, only a small number of studies have examined the influence of grandparents on grandchildren's values. The few existing studies are not entirely consistent, but they suggest that the influence of the oldest generation on the youngest generation is not very large, if it exists at all.[8] Our study is no exception. We discovered very little evidence of intergenerational transmission of attitudes from grandparents to grandchildren.

All three generations—the children and parents in 1981 and the grandparents in 1983—were asked to respond to a similar set of questions about family values that had been asked over the years in various public opinion surveys. They are displayed in table A–8 in appendix 4. As might be expected, there is some general shift toward less traditional family values with each descendant generation. The grandparents are somewhat less tolerant of divorce, of sexual permissiveness, and of working parents than are their children and grandchildren. Perhaps the sharpest difference is that just 6 percent of the grandparents agreed that "living together before marriage makes a lot of sense," compared to 23 percent of the parents and 42 percent of the grandchildren. But in general, the differences between the grandparents and the parents are larger and more consistent than the differences between the parents and the grandchildren. For example, 76 percent of the grandparents agreed that "marriages are better off when the husband works and the wife runs the home and cares for the children," compared to 58 percent of the parents and 55 percent of the grandchildren. Despite the general drift away from traditional standards, there is a fairly high degree of overlap in the family values of the three generations. Nearly everyone in the three generations agrees that "people should not get married unless they are deeply in love," and about three-fourths of each generation agrees that "unless a couple is prepared to stay together for life, they should not get married."

We used the statistical technique of factor analysis to ascertain whether certain subsets of these nine questions clustered together, indicating that they were measuring a similar dimension of family values. Two main clusters emerged, each consisting of three questions: the first (questions 77b, 77e, and 77g) seemed to tap the value respondents placed on the traditional family; the second (questions 77c, 77d, and 77i) measured their endorsement of contemporary family patterns. (Three questions were unrelated to either cluster and excluded from further analyses.) We calculated a score for each person on each cluster by adding up the number of "agree" responses to the three questions. Then, in order to assess the extent of intergenerational agreement, we calculated, for each scale, the correlation (a statistical measure of agreement) between the scores of grandparents and parents, grandparents and grandchildren, and parents and grandchildren. Table A–9 in appendix 4 presents these correlations, and the text of the appendix provides more detail about our methods.

The table shows statistically significant but fairly low levels of agreement on the two scales for each pair of generations. This means that grandparents, parents, and grandchildren in the same family share values to a greater extent than might be expected by chance, but that the degree of consensus within families is rather low. This pattern suggests relatively little transmission of values from one generation to the next. The level of agreement between the grandparents and grandchildren is somewhat lower than the agreement between the adjacent generations, but it is still greater than might be expected by chance. In other words, the attitudes of grandparents and grandchildren overlap more than would be the case if they were not in the same family, but the extent of influence of the oldest generation over the youngest is not substantial. The level of agreement between parents and children, though modest, is somewhat higher than between grandparents and parents. Surprisingly, adolescents seem to share their parents' values about the family more than adults share the values of their parents. It seems as though the more liberal family values of today are accepted by the two younger generations.

This is the general rule; but surely, we thought, there must be families in which grandparents have a more prominent role in shaping the views of their grandchildren. In order to see if we could identify such families, we carried our analysis one step further. We tried to identify circumstances that might increase the influence of the grandparents on their grandchildren. We reasoned that the level of agreement would be higher when grandparents lived near their grandchildren, when the senior generation placed a high importance on family relations (that is, had a high score on the family ritual index discussed in previous chapters), when they reported a high amount of exchange, when they took a greater parentlike role in childrearing, and when they were emotionally close to their grandchildren. All of these conditions, we thought, should promote greater intergenerational transmission of values from grandparents to grandchildren.

To test this hypothesis, we first computed the absolute level of agreement between grandparents and grandchildren on each of the two clusters of family attitudes. Then we used the statistical technique of multiple classification analysis to determine whether any of the aformentioned conditions increased the level of agreement on the two attitude measures.[9] They did not. Apparently, the nature of the relationship between grandparents and grandchildren does not affect the likelihood that they will hold similar family values. We also inspected a number of social characteristics of the grandparents and grandchildren to determine potential sources of influence. Characteristics such as race, family disruption, and proximity, which made a difference in the importance of grandparents to grandchildren, were nevertheless unrelated to intergenerational agreement. In sum, there is modest agreement between the generations, but it does not increase when relations between grandparents and grandchildren are more intimate or when involvement of the grandparents in the socialization process is greater.

One reason, we believe, for these generally negative findings is that grandparents do not directly transmit values to their grandchildren. Rather, the middle generation serves as the intermediary: grandparents have some influence over their own children, but they let the children do the work of guiding and influencing the grandchildren. In our personal interviews, several grandparents said that influencing the grandchildren's values was the children's job. One grandfather said:

> As far as influencing the grandchildren, I wouldn't even try, because I think my son and daughter-in-law are doing a wonderful job.

This view is consistent, of course, with the general reluctance of grandparents to get involved in childrearing decisions.

But even grandparents who are involved with their grandchildren and who do exercise authority over them seem to have limited influence over the grandchildren's values, as the following exchange suggests:

> INTERVIEWER: *Why don't grandparents have more influence over their grandchildren?*
>
> I raised [my grandson] for a year, so I feel closer to him than I do to the other children. I do have a little influence over him, but not much.
>
> INTERVIEWER: *Why not more?*
>
> My daughter now lives out of town. But when he was living in my house, I had the influence over him. . . . I don't see him every day. When I was seeing him every day, I was the one who was like raising him.
>
> INTERVIEWER: *What kind of influence did you have over him?*
>
> Let's see. Well, I would tell him to do something, and he would do it.
>
> INTERVIEWER: *So you had that kind of authority.*
>
> Yeah, I was like an authority figure.

> INTERVIEWER: *Do you think you ended up influencing the kind of person he is or the kind of values he has?*
>
> I don't know.

This grandparent is confident about the authority he wielded but quite uncertain about his influence in the realm of values. It appears that the transmission of values does not automatically follow from an involved relationship. Moreover, when asked about whether they influence their grandchildren, several grandparents discussed only the respect, or lack of it, they get—as if the value they were most concerned about passing along was respect for elders:

> INTERVIEWER: *Would you want to influence your grandchildren more?*
>
> I don't know. They more or less sort of respect us. I think my son and daughter-in-law, they, you know . . . we're important to them.

For many grandparents, then, influence seems to be defined as getting respect, making sure your grandchildren value you—rather than passing along moral codes to them. Perhaps grandparents have accepted this limited definition of influence because of the difficulty all family members face in shaping children's attitudes toward family life. Another grandfather complained:

> There are enormous outside influences which are taking over. The radio, the television, the movies, you send the kids away to college. . . . The minute the kids get infused into society, they lose—they don't know what a good, warm family life is because those influences that used to be confined around the family are now diffused.

The competition from the media, the peer group, the schools, and so forth can make it difficult for parents to shape their children's values; given that grandparents have less leverage over their grandchildren, the task is even tougher for them. Most, apparently, leave the transmission of values to the parents.

If our supposition is correct that grandparents directly transmit values to their children but not to their grandchildren, we would expect that the level of agreement between grandparents and grandchildren on the family values questions would be substantially reduced if we could remove the mediating influence of the parents. Statistically speaking, the correlation between the grandparents' and grandchildren's responses should fall to near zero once we control for the parents' responses. That is precisely what happened, as we show in table A–9 in appendix 4. When we controlled for the parents' values, the correlation between the grandparents and grandchildren virtually disappeared on both of the measures of family values.

The picture looks like this: the extent of grandparents' influence on their grandchildren's values is mediated through the middle generation. Generation One influences Generation Two, which in turn influences Generation Three. Grandparents pass on a legacy of values to their grandchildren if and only if they are successful in transmitting their values to their children. If this process of intergenerational transmission is interrupted, then grandparents will generally not exert any independent influence over their grandchildren's views.

Our results, by the way, do not give any credence to the popular perception that values skip a generation—that children rebel against their parents and, in doing so, often return to the values of their grandparents. On the contrary, we find that if there is a generation gap between grandparents and their adult children, it generally becomes wider between grandparents and their grandchildren. Conversely, when adult children and their parents share a common view of the world, at least about family values, the likelihood that grandchildren will share their grandparents' outlook is increased.

The Influence of Grandparents on Grandchildren

Our examination of intergenerational transmission has focused on only a single realm, values about the family. It is conceivable that consensus might be greater on other topics. But as we have pointed out, our findings are generally consistent with the scattered findings of other researchers who have explored the influence of grandparents on their grandchildren's attitudes and values. Most have reached the same conclusion we did. There is very little evidence that grandparents' values influence the values of their grandchildren. It seems to us that differences in age and life experiences often create different perspectives that more than offset the potential influence of common family membership. In addition, the competing influences of the outside world and the general reluctance of grandparents to interfere mitigate against the transmission of values. Teenagers and their grandparents may need to reach an understanding that they do not share the same world view.

The Influence of Grandparents on Their Grandchildren's Behavior

Before we totally dismiss the possibility that grandparents exert influence upon their grandchildren, we should consider one further test of our data. It has been argued by some observers that children benefit psychologically when grandparents play a more active role in their upbringing.[10] Participation by grandparents in the socialization process is thought by some to provide an important source of adult guidance, reinforcing the role of parents, especially when the child is living in a single-parent family.[11]

But as we noted earlier, few researchers have actually attempted to measure the effect of grandparent involvement on the development or well-being of grandchildren, and most of those have relied on clinical evidence with very small and selected samples. Perhaps the most compelling evidence of grandparent influence was supplied by Sheppard Kellam and his colleagues, who demonstrated that black children profited when grandparents played a more active role in their upbringing.[12] But it is difficult to know whether Kellam's results can be extended beyond the special circumstances of the low-income blacks in his study. Our study provides an opportunity to examine this question. In the National Survey of Children, a vast amount of data on the children's emotional and social behavior were collected from the parents, teachers, and the adolescents themselves. In previous analyses, a series of behavioral measures were developed, including independent measures of the child's antisocial behavior as reported by parents, teachers, and children; of psychological adjustment separately reported by parents and children; and of academic performance as reported by the teachers. A list of the scales can be found in appendix 4.

In previous research, we learned that children living in maritally stable families had different scores on these behavioral adjustment scales than did children from disrupted families. The scales also show clear behavioral differences according to variations in children's socioeconomic status, mother's age at first birth, child's gender, and other selected social and demographic characteristics of the family.[13] Thus, we have reason to believe that the scales do indeed reveal important features of the child's social and psychological functioning.

When grandparents are actively involved in the child's life, do children experience fewer behavioral problems and psychological difficulties? This hypothesis was tested in several ways by relating different indicators of grandparent involvement with the array of child adjustment measures. The results were remarkably consistent. Higher involvement by grandparents did not affect the child's social or psychological adjustment positively.

Specifically, first we looked at whether grandchildren of grandparents who reported that they were closer to and had more contact with their grandchildren actually had fewer problems at home and at school. They did not. None of the adjustment measures was significantly associated with the quality or intensity of the relationship between grandparents and their grandchildren. Next we examined whether the style of grandparenting had any detectable effect on the child's behavior. We expected that the grandchildren of more involved grandparents would have fewer behavioral and psychological problems. If anything, the opposite was true. According to several of the measures, the grandchildren of less involved grandparents experienced fewer adjustment problems, though most of the differences were not large enough to reach statistical significance.

We do not believe these results indicate that a more active relationship with grandparents has a negative effect on children. It seems far more plausible that when children are experiencing problems, grandparents are more likely to assist the parents. In earlier chapters, we discovered that grandparents are frequently called upon in times of family crisis. Marital troubles, illness, and bouts of unemployment are occasions when grandparents are called upon for assistance; they also are situations that frequently disturb the well-being of children. Our data suggest, then, that parents may be especially inclined to seek help from grandparents in raising their children if they are having difficulties handling them alone. Therefore, in families where grandparents are actively involved in childrearing, the grandchildren are not making out quite as well. But they probably would be doing even worse if the grandparents were not lending a helping hand.

Taking this possible scenario into account, we examined disrupted and nondisrupted families separately. Perhaps, we reasoned, more active participation by grandparents benefits grandchildren in nondisrupted families. Not so; our results again pointed to the conclusion that children do not generally make out any better when their grandparents are playing a more active role in their upbringing. Regardless of the family circumstances, higher levels of participation by grandparents have little effect on the child's overall adjustment.

The findings we have presented lead us to question the importance of grandparents in the socialization of children. To be sure, our conclusions must be tempered by the kind of data we have collected. The information produced by our survey does not reveal the subtle forms of influence that occur when grandparents and grandchildren interact over long periods of time; we are only looking at a small portion of what is obviously a very complex picture. Nonetheless, we believe that the data we have presented are persuasive in certain respects.

First of all, we discovered that only rarely do children consider grandparents to be primary support figures. A tiny fraction say that when problems arise, they are likely to turn first to a grandparent. Few children regard their grandparents as parentlike figures, and most do not include them when they are asked to list the people who make up their family. Even those who do so are not any more likely to regard their grandparents as persons to whom they can look for guidance in forming their social views.

We also found little support for the hypothesis that grandparents influence their grandchildren's values about family life. Such agreement as there was could be explained entirely by the linkage provided by the middle generation. When parents shared the grandparents' views, grandchildren and grandparents were more likely to hold similar views. When parents deviated from the grandparents' views, then the grandchildren did too. In short, there was no evidence that grandparents exerted a direct, independent influence on their grandchildren's attitudes and values.

Of course, we were only able to examine one realm of values. It is possible that greater agreement would be discovered in others. But our findings resemble the results of other investigators who have examined the extent of intergenerational continuity, leading us to place more confidence in our data.[14] Besides, it is certainly reasonable to expect that if the outlook of children were strongly affected by their grandparents' views, it would be evident in the area of family life.

Our final effort to trace the influence of grandparents on their grandchildren concerned the potential impact of the senior generation on the psychological well-being of the children. Do children experience fewer problems when they have more intense contact with grandparents? Using a number of different measures of child adjustment and several indicators of the relationship with grandparents, we could discover no evidence that greater involvement by grandparents had a positive effect on the behavior or mental health of the children in our study.

Indeed, if anything, the grandchildren of less involved grandparents seemed to have fewer behavioral problems. This is not, we believe, because grandchildren do worse when grandparents play an important role in their lives; rather, when children experience problems, grandparents are often called upon to help out. These findings are consistent with the data presented in earlier chapters that show that grandparents are expected to provide assistance in times of family crisis. They are the parents' reserves when things go wrong. In our case studies, we found several instances in which children who were experiencing problems were sent to their grandparents for temporary shelter.

Thus, the findings in this chapter are consistent with the picture provided throughout the volume. Grandparents are not prominent or visible actors in our family system, but they are important backstage figures. They are called on to act when problems arise. That is why we discovered that grandparents played a more active role in families where children were experiencing problems. And that is why we discovered that children in maritally disrupted families were more likely to mention their grandparents as family members than those in intact households. Grandparents in America are like volunteer firefighters: they are required to be on the scene when needed but otherwise keep their assistance in reserve.

To be sure, there are instances where grandparents play a pivotal part in a child's upbringing, such as when a divorce occurs and the grandparent substitutes for the absent parent. But, as we saw in the preceding chapter, such instances do not happen frequently enough to affect the general picture. As we observed earlier, grandparent-grandchild relations mirror parent-child relations. If parents have relatively little influence in shaping the values of their adolescent children, grandparents—who typically see their grandchildren only from time to time, largely on recreational or ceremonial occasions—are likely to have an even weaker influence.

Can we expect grandparents to have a different place in the family in the future? Could we rearrange family life to give grandparents greater influence over their descendents? In the next chapter, based on what we have learned in this study, we offer our best guesses about the future of grandparenthood in America.

8

The Future of Grandparenthood

Almost three years to the day after our initial visit, we returned to the senior citizen center where we had begun our study of American grandparents. During the interim, we had designed a questionnaire, conducted a nationwide survey by telephone, visited personally with grandparents around the country, analyzed our data, and written most of this book. Still, we wanted to listen to grandparents one last time, to test whether their words would now have a familiar ring, and to clear up a few lingering issues.

So, feeling now like veterans rather than the novices of 1982, we once again parked our car in front of the converted store and went inside to meet a new group of Jewish grandparents. There had been some changes in the neighborhood. In the past few years, several Jewish families had emigrated to this city from the Soviet Union. Our informants were in awe of the relationships between the Russian-Jewish grandparents and their families. One woman, Mrs. Berg, told us:

We have quite a few Russian friends. Now these friends,

invariably the mothers are living with the sons, the daughters, with grandchildren—they're all together. I have noticed such a difference among our Russian friends, who have only been in this country four or five years—these are new immigrants. The grandchildren have great reverence for the grandparents; they live together. When they go on vacation they take the grandparents with them! I'm just shocked! When they go out eating, they take the grandparents with them! When you go to a restaurant, you will see the grandparents with them. And this is such a new experience because we don't go eating with our children. They go their way, we go our way. It's a special occasion when you go out with your children—an anniversary, a birthday.

Her husband added:

The [Russian] grandparents are very much involved with the families of their children. They assume such a responsibility. We know a woman who was a famous surgeon in Russia, she's here in America, and she has that same feeling that she has to take care of her grandchild if the mother goes away. She's always obligated to that little girl. Now, you wouldn't find the same thing in America.

The grandparents in the room contrasted this great degree of togetherness, respect, and mutual obligation with the weaker ties they experienced with their own children and grandchildren. Mr. Berg asked:

Does it appear to you that we're disenchanted?

INTERVIEWER: *Do you think you're disenchanted?*

Yes. Disenchanted with the way we expected the family continuity to be—you know, when we were younger, when we had children, and the way it turned out. The lessening of the relationship, the distance between us, the wide abyss

> between the way my grandchildren think and the way we feel. And the kind of reverence we had toward an old, intelligent, crude grandfather. I had reverence for him because he was my grandfather.

After a number of these contrasts between the fullness of the grandparent-grandchild relationship among the Russian immigrants and the thinness among the Americans, plus some complaining about the material advantages immigrant families had secured through special assistance programs, one of us asked whether anyone in the group would trade places with the immigrants in order to have their type of relationship. The question was met with immediate cries of "No way!", "No," "I'm satisfied." The questioner pursued the point further: "Why wouldn't you trade places? There are all these strong family ties." A woman replied, "I don't think I could live with my children," and a chorus of "No" and "No way" followed.

Another woman said simply, "It's too late." And in an important sense, she was right. The immigrants' family relationships reminded the Jewish-American grandparents of the idealized picture that they conjure up when asked what family life was like when they were children. Whether or not this picture is accurate as a description of the past, it is clear to most grandparents that it cannot serve as a model for their families today—that, indeed, they would reject this model if it were offered. For the discussion at the senior center illustrated a central contradiction in the lives of American grandparents: like most other Americans, they want intimate, satisfying, stable family ties, but at the same time they want to retain their independence from kin. They want affection and respect from their children and grandchildren, but they do not want to be obligated to them. The price paid for strong family ties by the Russian immigrants—and by family members in developing countries around the world—is a substantial loss of autonomy. It is a price most American grandparents are not willing to pay.

In this regard, they are becoming more and more like their children and grandchildren. Joseph Veroff, Elizabeth Douvan, and Richard A. Kulka analyzed two national surveys about Americans' feelings of well-being and life satisfaction, conducted in 1957 and 1976. Overall, they found that feelings of well-being among Americans in 1976 were tied to personal growth more than in 1957. Life satisfaction was linked more closely to interpersonal intimacy and less closely to participation in organizations such as the church or social roles such as worker or husband. Moreover, self-reliance and self-expression became more important sources of fulfillment. And as for the greater importance of self-reliance:

> This is no more clearly highlighted than in the case of older people in the society of 1976 who have joined the rest of the population in seeking self-sufficiency as a crucial life value for well-being more than older people in 1957.[1]

We would argue that this change in the basis of older people's sense of well-being is rooted in the material changes discussed in chapter 2: the great rise in their standard of living, the increase in longevity, improvements in transportation and communication, and the like. As a result of these trends, many more older Americans have the opportunity to live independent lives. Given this opportunity, they—like their children and grandchildren—are seizing it. Another grandparent at the senior center told us:

> When we were raising children, we figured when the children got married and moved to their own locales, we'd be free to do as we please. All our lives, we've worked for the kids, to make sure they had an education and everything else. Then we find out when we're grandparents they say, "Uh,

> Mom, Pop, how about babysitting? We're going away for a couple of days." Once in a while this is fine, but we wouldn't want to be tied down to that like three or four times a year. And when my wife was babysitting for my granddaughter, I was against it because she was coming here to the center on Wednesdays. She was giving up her day; it killed the day even though she was only babysitting for five hours or so. . . . And she's depriving herself of her pleasures.

The sense that they deserve to have their pleasures now because they worked hard to raise their children was widespread among the grandparents in our study. Having paid their dues, many—like the grandfather just quoted—felt that they need not be "tied down" too often, need not oblige all their children's demands. Perhaps grandparents in other cultures, if faced with the same opportunities for independence, would make different choices. For example, older people in Taiwan still live with their adult children in substantial, though slowly declining, numbers despite rapid economic development.[2] But personal autonomy has long been a central value in American life, as observers from the time of Tocqueville have noted. And this emphasis on autonomy extends to family relations as well. In a recent study of American individualism, Robert N. Bellah and his colleagues wrote that "free choice in the family, which was already greater in Tocqueville's day than it had been before, is now characteristic of the decisions of all members of the family except the youngest children."[3] Given the central place of personal autonomy in American culture and the improved material circumstances of older people, the shift among grandparents toward greater independence seems inexorable.

We would also argue, along the lines stated in chapter 2, that the increasing independence of the generations promotes the growth of the companionate style of interaction between grandparents and grandchildren. Informal, affectionate, warm relations are more likely when grandparents and grandchildren are relatively equal in social status, as anthropological studies suggest.[4] Thus, the greater independence of grandparents—their reluctance to assume responsibility except in times of crisis, their exclusion by parents from decision-making, their overall lack of authority—leads to a greater emphasis on personal intimacy and emotional satisfaction with grandchildren. Here again, as the Veroff, Douvan, and Kulka study makes clear, their behavior mirrors that of Americans in general. Grandparents, newly freed from the constraints of economic dependence, blessed with longer lives, and imbued with American values, have joined their juniors in the pursuit of sentiment. The new American grandparent wants to be involved in her grandchildren's lives, but not at the cost of her autonomy.

Some would fault grandparents for this behavior. Arguing on behalf of stronger family ties, conservative critics of the contemporary family have called for a return to "traditional" family values, including the restoration of the authority of old over young and of husbands over wives. Although we respect the moral concern that underlies this position, our research has convinced us that the chances for a large-scale restoration of these traditional values are near zero—particularly insofar as grandparents are concerned. The most outspoken advocate for closer ties between grandparents and grandchildren has been child psychiatrist Arthur Kornhaber, whose views were introduced in chapter 2. He has warned of a "new social contract" that has, in his opinion, weakened the family:

> A great many grandparents have given up emotional attachments to their grandchildren. They have ceded the power to determine their grandparenting relationship to the grandchildren's parents and, in effect, have turned their backs on an entire generation.[5]

In the book he wrote with journalist Kenneth L. Woodward, he calls for a return to "an ethos which values emotions and emotional attachments," particularly between grandparents and grandchildren.[6]

Kornhaber's critique, though well intentioned and not without basis, is strong on rhetoric but weak on the facts. First, he downplays the positive features of social change. For example, his charge that grandparents have abandoned emotional attachments to their grandchildren is false, as the evidence in this book has amply demonstrated. Again and again, our interviews showed that grandparents have strong attachments to their grandchildren. In a majority of cases, this bond takes the form of a companionate relationship based on regular (though not daily) contact and an informal style of interaction. The grandparents in our survey reported overwhelmingly, as noted in chapter 2, that their relationships with the study children were "closer" and "more friendly" than their own relationships with their grandparents had been when they were children. We learned that grandparents with companionate relationships expressed great contentment with the emotional rewards of grandparenthood, even if they were not completely satisfied with the amount of time they were able to spend with their grandchildren. In chapter 5, we discovered that the major enemy of grandparents is geographical distance. When grandchildren live nearby, grandparents see them often, even if they do not get along with the children's parents.

Kornhaber, as we noted in chapter 2, does not think today's companionate relationships qualify as "real" relationships—thus his charge that grandparents have abandoned their family commitments. But they certainly felt like real relationships to the grandparents we spoke with. One should at least consider the possibility that the greater emphasis on companionship has had the salutary effect of diminishing the stiff, formal style that often dominated intergenerational relations in the past and increasing the salience of love and affection in intergenerational relations today.

Similarly, there is little recognition that the greater independence of the older population usually is experienced by them and most of their children as a positive development. What seems to Kornhaber as detachment is perceived by many grandparents as self-reliance, a quality much valued in American culture. Just a generation or two ago, far fewer grandparents had the economic resources and the good health necessary to live long, independent lives. Now that more do, is it fair to criticize them for enjoying the autonomy that is so highly prized in American society? The older among today's grandparents came of age during the Great Depression and lived through the hardships of World War II. They raised large families during the baby boom and worked to put their children through college. Consequently, it is not surprising that most older Americans are generally content with the trouble-free, independent lives that they now lead. Is there not a great social achievement here, namely, an advance in the quality of life for a previously disadvantaged group of Americans?

Conservative arguments about the family are often stated as if, without any major changes in social structure, moral exhortation alone could alter people's behavior patterns. Kornhaber and Woodward, as we discussed in chapter 2, are aware that the strong intergenerational ties of the past that they so admire were rooted in economic cooperation among families working hard to subsist. They further acknowledge that as material conditions have improved, intergenerational ties have become less intense. Yet they reject the linkage between material conditions and family relationships, calling merely for the spread of an ethos than would restore the intensive emotional ties of the idealized three-generational unit, as if the power of moral suasion were sufficient to bring about the changes they favor. It is not. To attain the goal they seek, one must consider, in addition, what changes in the social structure would be necessary —and what the social and economic costs would be.

Let us consider, then, what it might take to establish widely the kind of deeper, stronger ties between grandparents and grandchildren that Kornhaber—and, other things being equal, almost everyone else—would like to see. To begin with, our society would have to discourage if not restrict people's geographical movements. The analyses in chapter 5 showed that involved relationships between grandparents and grandchildren depend most heavily on very frequent contact, which in turn depends very heavily on living nearby. A massive relocation would be required to put enough grandparents and grandchildren near enough to allow for a great increase in contact. Presumably, young adults move to pursue better opportunities, thus improving the general welfare (by more efficiently matching their skills to jobs); and older people relocate for health and recreational reasons. Are we prepared to reduce employment opportunities for improved grandparent-grandchild relations? Would we really consider urging workers to remain in depressed labor markets so that their children would have regular contact with grandparents? How many older people who have worked hard all their lives could be dissuaded from retiring to a condominium in the Sunbelt? We doubt that many Americans would respond to these appeals.

Moreover, even if such appeals were successful, they would not suffice. The involvement of grandparents in families of the past was enhanced when they retained a substantial amount of authority over the lives of their children, often based upon the ownership or control of economic resources. Even today in many developing countries, elders have a great deal of influence over the timing of the major events in their children's lives due to their control over resources.[7] In order to foster intense intergenerational ties, then, our society would have to give the older generation substantial control over resources and allow for older people to have more influence over their children's choice of spouse, timing of marriage, type of work, and place of residence. We suspect that few Americans would accept this degree of influence in their own family decisions.

The Gains from Intergenerational Ties

These, then, are the steps it would take to restore the authority of grandparents as it is sometimes believed to have existed in the past. We stress the word "believed" because family historians have cast doubt on the notion that American families ever functioned like families in Japan, India, or China, where reverence and respect for ancestors remains high even today. Still, it is likely that family elders more often wielded influence in day-to-day family dealings a century ago than they do today. Teenaged children were expected to contribute to their parents' support, and elderly parents often continued to receive material support from their adult children and their children's children. Money and services flowed in two directions within the family —to children when they were young and then back to parents as they aged.

Family scholars are not quite certain just when and why this reciprocal flow of resources changed. Some believe that the change occurred gradually as parents lost the ability to employ their children in the family farm or business. Others think that the sustained period of economic growth following World War II reduced the need for intergenerational transfers. Probably, too, the Social Security system, which was devised to insure the economic well-being of the elderly, helped to undermine the traditional pattern of intergenerational exchange. By the middle of this century, relatively few adult children continued to turn over portions of their earnings to their parents. Increasingly, older parents were able to support themselves, and growing numbers of them were able to assist their adult children in establishing careers and launching families.

What do grandparents gain from intergenerational relations today? First, as we observed in chapter 2, an indirect exchange still is taking place through the Social Security system. The middle generation is paying generously to support the elderly, while preserving the cultural fiction that the elderly are taking care of themselves. This belief reinforces the high value Americans place on the autonomy of family members. Grandparents do not want to rely on their children and grandchildren for economic security.

Second, grandparents are repaid by their children and grandchildren in sentimental currency—love and affection. While we cannot fully document it, we believe that this form of exchange probably has increased over the past several decades. We have argued that emotional ties between the elderly and their grandchildren are more valued than ever before. This does not necessarily mean that they are more gratifying—though we suspect that they often are—but that the standards of what constitutes a "close" relationship have been elevated. As in contemporary marriage, more emotional satisfaction is now expected in intergenerational relations. Yet, as in marriage, the pursuit of sentiment, though personally rewarding, can be elusive and frustrating. An inflation in standards can lead to doubts and disappointments. When our respondents told us about their strong feelings for their grandchildren, we sometimes sensed reservations about whether the feelings were reciprocated. Some grandparents felt less than secure about their place in the family.

Third, grandparents are able to bask in what they perceive as the special achievements of their grandchildren. Though they may see them infrequently, grandparents boast of their grandchildren's success in school, on the athletic field, or in church activities. In the homes we visited, we were shown trophies, merit badges, certificates, and newspaper clippings. We were regaled with stories of school plays, honor rolls, and lucrative summer employment. It was clear that grandchildren's accomplishments had helped to make life worthwhile. To be sure, grandparents understand, as do parents, that children are achieving for themselves. But grandparents are a central part of the audience. Indeed, parents encourage them to share the children's accomplishments. And one of the benefits of this form of exchange is that grandparents who live in Florida or Arizona can receive it nearly as well as those who live next door.

Our conclusion, then, is that there will be no "return" to "traditional" grandparent-grandchild relations on a large scale —and that this is not altogether bad. Substantial benefits have accrued to grandparents and their adult children as a result of the movement toward greater autonomy and companionship. Moreover, the symbolic rewards of grandparenthood are important, even though they are less substantive than the exchange of goods and services. In addition, the costs of widely establishing a strong, influential relationship between grandparents and their progeny are so high that few would support the necessary actions. This is not to deny that real costs are attached to the current state of family relations or that some grandparents have misgivings or discontents about their roles. All family arrangements entail certain costs; we are probably more aware of the costs of ours than ever before. Rather, we wish to emphasize here that even if there are structural weaknesses in the American family (as few would deny), grandparents are not likely to be the agents of change or useful instruments of public policy. Their influence in most families is modest, and the prospects for greatly increasing that influence are slim. Those searching for a means of strengthening the American family will probably have to look elsewhere.

Countercurrents

Still, there are some countervailing tendencies that may act to increase modestly the importance of the grandparental role. Those who are inclined to exhortations on behalf of strong bonds between grandparents and grandchildren might take heart from our finding that grandparents with a greater family consciousness saw their grandchildren more often and exchanged more services. These findings showed that there is significant variation in family values among American grandparents, holding economic factors constant. But just how one might foster the spread of a more familistic orientation—assuming that was a shared policy objective—is unclear.

More important, for better or worse, is the trend toward frequent divorce. We have seen how a divorce in the middle generation creates a need for assistance from the older generation and how grandparents, particularly on the custodial side, respond to that need. The crisis of divorce calls into action the latent support network of the family, in which grandparents play a central part. In essence, divorce recreates a functional role for grandparents similar to the roles they had when higher parental mortality and lower standards of living necessitated more intergenerational assistance. One result, as was shown in chapter 6, is that children in divorced families today tend to develop stronger ties to their custodial grandparents than children in intact families develop with either set of grandparents.

The effects of divorce demonstrate once again that strong, functional intergenerational ties are linked to family crises, low incomes, and instability rather than to health, prosperity, and stability. There is a tradeoff here, and we suspect that almost everyone would prefer to see fewer family crises—and less divorce—even at the cost of a weakening of intergenerational ties. Consequently, the latent nature of intergenerational support in most intact families today—the "family watchdog" role of grandparents, in Lillian Troll's phrase—should be seen as an advance in social welfare. It reflects the greater prosperity and (at least until the recent rise in divorce) stability of the nuclear unit and the greater material well-being of the older generation. It would be nice to have one's cake and eat it, too; that is, to have prosperity and family stability and also to retain strong intergenerational ties. But the strong bonds of the past, when they did exist, derived from day-to-day participation in a common family enterprise. Without the need for such participation, intergenerational ties emphasize loving, affectionate, companionate relations with a fund of additional resources held in reserve.

Nevertheless, divorce is creating a more functional role for grandparents in millions of American families. At current rates, about two-fifths of all American children will witness their parents' divorce, as noted in chapter 6. Since most older people have more than one child, the odds are increasingly high that a grandparent will watch at least one child divorce after the birth of grandchildren. The small silver lining in this otherwise dark cloud is that many grandchildren will experience closer ties to some of their grandparents. But as was shown in chapter 6, this benefit is distributed quite unequally, from the grandparents' point of view. Maternal grandparents, whose daughters usually retain custody following a divorce, stand to benefit most; but paternal grandparents stand to lose. Although some paternal, noncustodial grandparents do establish closer ties, they more commonly become symbolic figures who see their grandchildren infrequently and on ritual occasions. Nor, as we argued, is this situation amenable to legal remedies. High rates of divorce, coupled with current custody patterns, could increase the relative importance of the maternal line in American kinship, an unanticipated, unwanted, though not necessarily harmful effect. All in all, divorce is far from an ideal way to strengthen intergenerational ties.

Another potential crosscurrent is the increasing relative wealth of older Americans. Because of the economic trends discussed in chapter 2, the elderly are now a relatively advantaged, though not affluent, group. Should these trends continue, more grandparents will be able to play the role of economic protectors of their grandchildren. They also may be more motivated to do so, as we will discuss, if birth rates remain low and grandchildren are in short supply. The family watchdogs will have more resources with which to act when trouble arises. Even so, we would not expect grandparents in intact families to have more authority over the major decisions in their children's lives. That kind of authority cannot be bought; it requires day-to-day involvement in the grandchildren's lives. Many of today's independent grandparents do not want this type of involvement, nor do their children want them to have it. And some of the grandparents who would like day-to-day involvement live too far away to get it. The typical types of financial aid are likely to be lump-sum transfers for special needs—such as orthodontics, college tuition, or a downpayment on a home—rather than smaller, regular payments. Indeed, only one-fourth of the most affluent grandparents in our survey (those with total family incomes of twenty thousand dollars or more in 1982) reported that they had provided any financial support to the study child's parents in the previous twelve months; and more than half of them did so "occasionally" or "seldom" (as opposed to "regularly").

Thus, grandparents are likely to be increasingly important as a source of financial reserves—an insurance policy against family crises or a source of assistance for major purchases—but not as a source of regular income support. This role is consistent with the independence both grandparents and their adult children value so highly. Occasional large transfers of assets allow grandparents to make an important contribution while retaining most of their regular flow of income for their own use. Occasional transfers also allow adult children to receive valuable assistance without feeling that they are dependent on their parents for support. And by compartmentalizing financial transfers, this system does not interfere with the preferred day-to-day emphasis on companionship.

On a national scale, one might ask whether these intrafamily transfers are likely to redress the growing imbalance between the well-being of children and the well-being of the elderly.

Samuel H. Preston has demonstrated that during the 1960s and 1970s, the relative economic position of American children declined while the relative economic position of the elderly improved. In 1970, according to Preston, 16 percent of children under fourteen were living in poverty, compared to 24 percent of persons over sixty-five. But by 1982 the situation had reversed: 23 percent of children were poor, compared to just 15 percent of the elderly. When in-kind payments such as Medicare and food stamps are taken into account, according to Preston, the gap widens further: 17 percent of children were poor in 1982, compared to just 4 percent of the elderly.[8] The gap is a result, in part, of the great increases in Social Security and pension payments noted in chapter 2. It also results from the growth of single-parent families and from the greater postponement of childbearing among more well-to-do families. Can we rely on private transfers from grandparents to grandchildren to help close the gap?

In our opinion, private, intrafamily transfers can be of only limited help. To be sure, moderately affluent grandparents whose children divorce often will be called upon to help out financially. We showed in chapter 6 that grandparents on the custodial side after a divorce were more likely to be providing support than other grandparents, even several years after the divorce. But the problem is that many disadvantaged children have disadvantaged grandparents. Out-of-wedlock births to teenagers, for example, occur disproportionately in the lower-income segment of the population; in some cases, the paternal grandparents may not even acknowledge (or know about) their son's paternity. In order to reduce substantially the disparity between children and the elderly, grandparents would have to take responsibility for assisting other people's grandchildren whom they do not know and who do not live in their neighborhoods. There is no reason to believe that this will occur on an individual level, not because grandparents are selfish but because, like most other Americans, they confine their personal generosity to their own families and their own communities. Even a sudden plunge in the divorce rate would not eliminate the gap. A fifteen-year study of family economics by the University of Michigan showed that only part of the reduction of children's economic well-being relative to the elderly is a result of changes in family structure.[9] Consequently, the only way to reduce the gap directly is by some type of aggregate transfer from the elderly—or from the even better-off adult population of working age—to children. Policy makers cannot look to private intergenerational transfers to mitigate the poverty that disadvantaged children face.

Grandparents: Supply and Demand

An additional trend that may have a substantial impact on the future of grandparenthood is the sharp decline in the birth rate since the 1950s. At current rates, the average American woman will give birth to fewer than two children, and continued low birth rates will create a growing imbalance between the numbers of grandparents and grandchildren. To illustrate, suppose we had done our survey in 1900, when the birth rate was higher and life expectancy was lower. As we noted in chapter 2, we would have found that grandparents were in short supply: there were only twenty-seven persons age fifty-five and over for every one hundred children fourteen and under. Moreover, those grandparents who were fortunate enough to have survived were less affluent and had fewer resources to give to their grandchildren. Now let us suppose that we were to redo our survey in the year 2000, and let us assume further that the birth rate has remained low, that gains in adult life expectancy have continued, and the economic situation of the elderly has not deteriorated. We would find that the demand for grandchildren would have increased as more persons lived relatively affluent, long lives but that the supply of grandchildren would have decreased sharply. To repeat again what we noted in chapter 2, the Bureau of the Census predicts that by the year 2000 persons fifty-five and over will actually outnumber children fourteen and under.

Thus, the demographic and economic bases of the grandparent-grandchild relationship at the end of this century are likely to be reversed from what existed at the beginning of the century. In the earlier period, grandparents had more claimants on their emotional and material resources and fewer resources to give; in the near future they will have more resources but far fewer claimants. A few generations ago some grandparents must have been overwhelmed by the number of grandchildren they had, but in the 1990s many more will be underwhelmed. On average, given continued low fertility, an older person will have nearly two living adult children, each of whom will have nearly two children. But these averages will conceal important variations. For example, demographer Charles F. Westoff estimates that if current rates continue, about one-fourth of all young women will not bear children.[10] If so, then a substantial minority of older persons will find that only one of their children will give birth to grandchildren—often just one or two.

This short supply of grandchildren may alter the strategies that grandparents pursue. Fewer grandparents will have the option of letting circumstances—such as geographical proximity or how well they get along with their daughters-in-law—determine the nature of their relationships with their grandchildren. The strategy we labeled selective investment in chapter 4 will not be possible as often. Instead, there will be more incentive for grandparents to invest heavily in their first or second grandchild, on the theory that there may not be any others, even if the grandchild does not live nearby. Yet it is very difficult for grandparents to overcome the barriers of distance or a poor relationship with daughters-in-law. There could be an increasing proportion of grandparents who have remote relationships with all their grandchildren. But if older people remain relatively well-off economically and fertility remains low, we would expect to see more grandparents engage in large expenditures such as airplane trips and joint vacations that might maintain ties despite great distances. Grandparents may also become even more accommodating in their relations with their children—even more circumspect, for example, about the norm of noninterference—in order to maintain family ties. Another alternative would be for grandparents to compensate by embracing stepgrandchildren as the best available substitute. We described in chapter 6 how the bonds between stepgrandparents and stepgrandchildren appear to vary sharply according to custody arrangements and to the age of the children when their parents remarry. Continued low fertility may promote the assimilation into the family of stepgrandchildren who today might be considered marginal.

All this suggests that the fit between the number of older family members, the resources available to them, and the numbers of children in their families—between the demand for grandchildren and the supply of same—may have improved and then worsened during this century. At the turn of the century, when grandparents were less numerous and less well-off, the demand for grandparents who could devote time, energy, and resources to their grandchildren may have exceeded the supply. By the end of the century the supply of able, healthy, well-off potential grandparents is likely to exceed the demand. We would speculate that sometime in the very recent past, probably in the 1970s, the balance between the resources and desires of grandparents and the needs of grandchildren may have been at a peak. The grandparents of the 1970s were the first to benefit from the extraordinary increases in Social Security payments that began in the mid-1960s. They also had large numbers of grandchildren because they had given birth to the large baby-boom cohorts in the 1950s. This conjuncture of longevity, relative affluence, and large families may prove to be unique. To a degree, it allowed grandparents to increase their personal autonomy without sacrificing intimacy—to lead independent lives without foregoing intergenerational ties. From the grandchildren's perspective, to be sure, the trends look different: they may benefit in the near future from the increased attention their scarcity seems likely to bring. Thus, a low birthrate, like a high divorce rate, may act to increase somewhat the salience of grandparents to grandchildren's lives. But grandparents as a group, it seems to us, never had it as good as they did in the recent past, and they may never have it as good again.

Formulating a New Definition of the Role

Finally, then, what are we to make of the ambiguous situation of the new American grandparent? First, we must reject the notion that grandparenthood is a meaningless, unimportant role. On the contrary, being a grandparent is deeply meaningful. But it must be understood that what makes life experiences "meaningful" in our society has changed in recent decades. Like most other Americans, grandparents increasingly find meaning in their lives through personal fulfillment: they seek self-reliance, and they seek emotionally satisfying interpersonal relationships. Their relationships with grandchildren fit these criteria. Eschewing the role of authority figure, grandparents concentrate instead on developing relationships based largely on love and affection. Most succeed in establishing rewarding, companionate relationships without compromising the autonomy they also value. Some social commentators argue that we ought to emphasize additional criteria when evaluating our personal well-being—for example, the extent to which we are engaged in efforts to assist others, or the degree to which we have lasting, stable bonds to family and community. But these arguments, valid though they may be, cannot change the fact that grandparents, in their own terms, find companionate relationships meaningful and satisfying. Whether or not it ought to be so, the pursuit of sentiment is central to the meaning of grandparenthood today.

Nevertheless, some grandparents play an active role in helping to rear their grandchildren and in exchanging services with them. Sixteen percent of the grandparents in our survey had this type of "involved" relationship with the study children. Since the grandchildren in the study essentially were selected randomly, an additional fraction of the grandparents must have had involved relationships with other grandchildren not in the study. In order for grandparents to have this functional role, however, a special set of circumstances must exist. Most important, the grandchildren must live quite close by; our findings suggest that, with few exceptions, grandparents cannot take on parentlike authority unless they see their children and grandchildren very frequently. It also helps if the grandparents are younger—hence better able to cope physically with the duties of caring for children—and if they get along well with the grandchild's mother. And it helps if there is a specific need for the grandparents' assistance, as is the case when a divorce occurs in the middle generation. There is not much chance of an involved relationship developing between a sixty-eight-year-old grandmother and a granddaughter who lives with her mother and father an hour's drive away; too many structural constraints prevent it, even if the grandmother wanted it. Thus, it is unrealistic to expect that large numbers of grandparents could play a strong, authoritative, ever-present role in their grandchildren's upbringing.

Is this to be lamented? Readers must answer that question according to their own moral views. But let us make a few observations that, we believe, should temper any feelings of disappointment about the infrequency of strong, functional intergenerational ties. First, as we argued in detail in chapter 2, the idealized picture of the strong, supportive grandparent of the past is overdrawn. As recently as the turn of the century, far fewer grandchildren had the opportunity to know their grandparents, and most did not live with them. It is misleading to compare the reality of the present to a nostalgic image of the past. Second, it is common for grandparents today to serve as the protectors of their grandchildren—as sources of support in reserve. This latent support never may be activated; but, like a good insurance policy, it is important nevertheless. In fact, given the growing affluence of the elderly, the rise in divorce, and the declining numbers of grandchildren, the protector role is likely to become more salient in the future. Third, the evidence in chapter 7 suggests that even if more grandparents played a key functional role, they would have a relatively modest influence over the values of their grandchildren. When even parents have a difficult time combating the influence of the media or the peer group, we cannot expect grandparents to be very effective.

What has happened over the past decades is that grandparents have been swept up in the same social changes that have altered the other major family relationships—wife and husband, parent and child. The increasing economic independence, the greater emphasis on self-reliance, and the search for sentiment that have changed marital relations—making them both more oriented toward intimacy and more brittle—have also changed grandparental relations. In addition, we have seen how the limitations of grandparent-grandchild relations mirror in many ways the difficulties of parent-child relations. Indeed, the final lesson we wish to draw from our study is that the situation of grandparents today demonstrates the pervasiveness of family change. Rooted in our changing social structure and our changing values, the tension between personal autonomy and family bonds now affects all important family relationships. For grandparents, this tension means that they must try to balance their desire for emotionally satisfying relationships with their grandchildren, on the one hand, against their wish to lead, at long last, independent lives. The resolution of this tension is the fundamental problem facing American families today. As our study suggests, there can be no return to the "traditional" family values of the past without alterations in our social structure that few would tolerate. And yet the maintenance of family ties remains important to most of us. This is a problem that requires the attention of all family members and of our society as a whole; we canot expect grandparents, acting alone, to solve it for us.

Appendices

Appendix 1

The Grandparent Study

This appendix provides a detailed description of our national survey of 510 grandparents conducted in 1983. The appendix has three sections: an explanation of the survey procedures, an analysis of the representativeness of the sample of grandparents, and a reproduction of the questionnaire used in the survey.

Survey Procedures

Our study was an extension of a continuing project first begun in 1976, when the Institute for Survey Research (ISR) at Temple University, under the direction of Nicholas Zill and with financial support from the Foundation for Child Development, carried out a national survey of children aged seven to eleven and one of their parents, usually their mothers. The purpose of the survey was to assess the well-being of American children. In 1981, Zill and Frank F. Furstenberg, Jr., with additional support from the National Institute of Mental Health, directed a second wave of interviews focused on the effects of marital disruption on children. ISR reinterviewed all of the children whose parents had experienced marital disruption by the time of the original 1976 interview, and a subsample of those in nondisrupted homes. As in 1976, one parent, usually the mother, also was interviewed. In anticipation of the grandparent study, the 1981 interviews with the parent included the following question:

> As part of our study, we would also like to conduct a telephone interview with [CHILD]'s grandparents with your permission. May I have their names, addresses, and phone numbers?

The respondent, if she was willing to do so, then provided the names, addresses, and telephone numbers of the grandparents on her side, her current husband's side (if she was currently married), and her ex-husband's side (if she was currently or previously divorced).

At the time of the 1981 interviews, 623 parents provided the names of 910 grandparents. Addresses were given for almost all of them and, in most cases, telephone numbers were also provided. Thirty-one percent of the 1981 interviews were with families in which two children had been interviewed in 1976 and 1981. Asking a grandparent about more than one grandchild would increase the length of the interview; moreover, we felt it might tax the respondent. We therefore decided that only one child would be designated as the "study child," the focus of the interview, and that child would be selected alternately, with the child listed first in 1981 selected every other time.

Because of budgetary constraints, we could conduct only about 500 interviews, so our next task was to reduce the sample size from 910 cases to about 625 cases. (We anticipated an 80 percent response rate, which would yield 500 completed interviews.) This was accomplished by deleting:

1. Grandparents from families in which the study children

were not living with at least one parent at the time of the 1981 interview (we wanted to focus our attention on grandparent-parent-grandchild linkages);

2. Grandparents of study children who were born before May 1, 1965, and were therefore eighteen years old or older by the time of the grandparent interviews in 1983;
3. Grandparents from families in which an interview with either the child or the parent was not completed in 1981;
4. Grandparents for whom phone numbers were either not provided by parents or could not be obtained by ISR; and
5. Grandparents who lived outside the coterminous United States.

After these deletions, 787 cases remained. In order to reduce this number to 625, the cases were subsampled in the following manner:

1. Children were divided into two groups. The first was composed of those whose parents had separated or divorced by the time of the 1981 interview. This group was given the designation SEP/DIV. The second group was composed of children whose parents were not separated or divorced by the time of the 1981 interview. The designation of NO SEP/DIV was given to this group.
2. All of the grandparents of children in the SEP/DIV group were kept in the sample. The number of grandparents (203) in the SEP/DIV group was subtracted from 787.
3. The remaining 584 grandparents came from the 399 children in the NO SEP/DIV group, whose parents had provided an average of 1.46 grandparent names and addresses per child. We needed to delete 162 of these grandparents to reduce the total sample to 625. We therefore deleted randomly 111 (162 divided by 1.46) children from the NO SEP/DIV group.

4. After these steps, 203 grandparents from the SEP/DIV group and 419 grandparents from the NO SEP/DIV (a total of 622 grandparents) remained in the sample.

At the time of the 1981 interview, parents gave the names of grandmothers 58 percent of the time, grandfathers 13 percent of the time, both 28 percent of the time, and only first initials 1 percent of the time. We wanted to ensure the inclusion of a substantial number of grandfathers, both widowers and the currently married. So, regardless of which grandparent was named by a parent, one-fifth of the cases were designated randomly as "grandfather" cases. Labels on the questionnaires for such cases instructed the interviewer to interview the grandfather if possible. On the remaining four-fifths, interviewers were instructed to interview the grandmothers if possible. In this way, 498 grandmothers and 124 grandfathers were designated for interviews.

Two administrative pretests were conducted. The objectives of these pretests were: (1) to assess whether a fairly lengthy telephone interview was feasible with grandparents, (2) to learn which of the questions were unclear or difficult to answer, and (3) to reduce the length of the interview to an average of no more than forty minutes. The first pretest was conducted from December 10 to December 17, 1982. Twelve interviews were completed, ten by two interviewers and one each by the Study Director and the Field Administrator at ISR. Most interviews were conducted with grandparents whose children had experienced marital disruption. (The subjects in the pretests were not taken from our sample.) Based on the debriefing discussion among Cherlin, Furstenberg, research associate Cheryl Allyn Miller, ISR Study Director Ellin Spector, ISR Senior Field Administrator Phylis G. Aronovitz, and the two interviewers, several revisions were made. From January 11 to January 18, 1983, the second pretest was conducted. A diverse group of respondents including natural grandparents, stepgrandparents, those whose children had experienced marital disruption, and those whose children were in intact marriages were interviewed by two interviewers. There was consensus that with minor changes the questionnaire would be easily administered and well received. We also determined that the average length of an interview would be approximately forty minutes.

In January 1983, twenty experienced interviewers were selected for the survey by ISR, of whom sixteen had interviewed children and parents in the 1981 study. Home-study packets containing the Interviewer's Instruction Manual (written by Cherlin, Furstenberg, and the ISR staff), grandparent questionnaire, and a quiz were mailed to interviewers. They were asked to become familiar with the manual and questionnaire and to complete the open-book quiz. Then they were required to conduct a practice telephone interview with a grandparent of an eight- to seventeen-year-old grandchild whose parents had been divorced since the child's birth or who were currently separated. The quiz and questionnaire were to be completed and mailed to ISR within a week. Practice interviews and quizzes with corrections and comments were returned to interviewers who were then authorized to interview grandparents from the sample.

Interviewers were assigned between fifteen and fifty-nine cases; an average assignment was thirty-one cases. They received Screening/Call Report Forms for each case and a printout with all of their cases. Each Screening/Call Report Form had a label with the case number (the same one that was used in the 1976 and 1981 interviews), the telephone number, the designation of the grandmother or grandfather as respondent, the name and address of the grandparent, the name of the study child, and the parent's marital status (SEP/DIV or NO SEP/DIV).

On February 8, ISR sent to each of the 622 grandparents an introductory letter explaining the purpose of the study and informing the grandparent that an interviewer would be calling to conduct a telephone interview. Enclosed with the letter was an update form that requested verification or correction of addresses and telephone numbers. Grandparents were also asked to indicate the best days and times for an interviewer to call. The ISR also sent to each of the 442 parents who provided the 622 grandparents' names a letter explaining the purpose of the grandparent study. Enclosed was a summary report of the 1981 findings and an update form with a request to verify or correct the grandparent's address and telephone number.

For such reasons as "insufficient address," "moved," "no forwarding address," "unknown address," and "no such address," the Postal Service returned forty letters addressed to parents and forty-four addressed to grandparents. For each undelivered letter, ISR attempted to find a correct address. ISR also received numerous grandparent and parent update forms, some of which contained different addresses and telephone numbers. Information on update forms also informed ISR that eleven grandparents had died, four were too ill to be interviewed, and six parents refused to allow grandparents to be contacted.

All update forms and post office corrections were sent to the interviewers as they were received. Interviewers were also provided with information ISR obtained from telephone directories and directory assistance operators. For cases that still could not be contacted, interviewers were asked to consult directories and operators again and, additionally, to call all others in that locality who had the same last name as the grandparent. Cases that could not be located by interviewers were returned to ISR, where a search was made of the 1981 interviews. Parents were called and asked for updated grandparent information and, for the most part, they were cooperative.

Each time ISR traced a grandparent, an introductory letter and update form was sent. In addition, the ISR Study Director wrote personal notes to all grandparents who did not have telephones or who had unlisted numbers. In response to these notes, two grandparents called ISR and were interviewed by the Study Director. Calls were also made to parents for current phone numbers of grandparents who the interviewers found to be away for the duration of the study. Some numbers for vacationing and traveling respondents were obtained that way. The result of all these tracing efforts is that interviews were conducted with fifty-five grandparents whose addresses had changed since 1981, twenty-five grandparents whose telephone numbers had changed since 1981, and thirty-four grandparents whose telephone numbers and addresses had changed since 1981.

All 510 interviews were conducted on the telephone between February and April 1983. If, for such reasons as death, illness, a language barrier, or a refusal, the grandparent designated on the label could not be interviewed, the other grandparent in the household was interviewed.

Grandparents in eight states around the country were asked at the end of the questionnaire if they would participate in a follow-up personal interview. Of the 145 grandparents who were asked, 112 agreed. In the spring and summer of 1984, thirty-six follow-up personal interviews were conducted in the grandparents' homes, the majority by research associate Miller. These interviews were tape recorded and subsequently transcribed.

Of the 622 cases in the sample, 510 were interviewed, an 82 percent response rate. The response rate was 80 percent for the SEP/DIV group and 83 percent for the NO SEP/DIV group. Interviews were completed with 446 grandmothers and sixty-four grandfathers. Forty-four grandmothers were interviewed when it was not possible to interview designated grandfathers. Nineteen grandfathers were interviewed instead of designated grandmothers. Two interviews were conducted in foreign languages (German and Spanish). The mean length of an interview was 35.7 minutes, with a range of 12 to 98 minutes. The typical time was about 30 minutes. Longer interviews were almost always due to some physical problem, such as with those who were hard-of-hearing. The disposition of the total sample was as follows:

Completed interviews	510
Unlisted phone	1
Respondent refused	32
Someone else refused	6
Respondent too ill	20
Language barrier	2
Respondent deceased	13
Respondent not child's grandparent	2
Respondent away for duration of study	4
Wrong number/no number	17
Respondent not home	2
No answer	4
Other (mainly partial interviews)	9
	622

If cases that could not possibly be interviewed (respondent deceased, not child's grandparent, or too ill) are deleted from the base, there are 588 cases and the response rate is 86.9 percent.

Representativeness

As we noted in the text of chapter 1, our sample is not, strictly speaking, a nationally representative sample of grandparents; rather, it consists of the grandparents of a nationally representative sample of teenaged children. Two biases are built into our sample by the very nature of our procedure. First, the grandparents in our sample had to have at least one grandchild aged thirteen to seventeen (the age range of the study children in 1983); therefore, grandparents with only younger grandchildren were underrepresented. Second, grandparents with many grandchildren had a greater probability that one of their grandchildren would have been selected for the 1976 study; as a result, grandparents with fewer grandchildren were underrepresented. These limitations must be kept in mind; it is possible that the experience of grandparenting differs for those with only young grandchildren or those with few grandchildren.

More important, however, are the potential biases introduced by our procedure for obtaining the names of the grandparents. As described previously, we obtained these names from the parents interviewed in 1981. We have evidence to suggest that some parents were more likely to provide us with names that led to completed interviews than were others. Specifically, parents who were better educated, white, less religious, or who had at least one parent who lived within an hour's drive or who had received assistance recently were more likely to have given us names that led to interviews. Thus, our sample of grandparents most likely overrepresents those who live nearby, provide assistance, are white, or are better educated. Still, as we will argue, the effects of these biases on our findings are likely to be modest.

In table A–1 we compare the characteristics of parents in those families in which we were able to obtain at least one grandparent interview with the characteristics of parents in those families in which no interviews were obtained. The five comparisons in the table all revealed statistically significant differences. (And all five remained significant in a multivariate analysis that was performed.) Regarding panel *A*, the parents in 1981 were asked if any of their parents or parents-in-law (the study children's grandparents) lived within an hour's drive. As can be seen, one or two grandparent interviews were obtained in 42 percent of the families in which the parent replied affirmatively, compared to 27 percent in which the parent replied

TABLE A–1

Characteristics of Those Families with Whom at Least One Interview was Obtained Compared to Those with Whom No Interview was Obtained

Number of Interviews Obtained in 1983	**Selected Characteristics of the Parents and Grandparents in 1981**		
	A. Proximity		
	No Grandparents Live Within an Hour's Drive	**One or More Sets of Grandparents Live Within an Hour's Drive**	
0	74	58	
1	17	27	
2	10	15	
	101%	100%	
	(390)*	(699)*	
	B. Assistance		
	No Assistance Received from Grandparents	**Assistance Received from One or More Sets of Grandparents**	
0	72	55	
1	20	27	
2	8	18	
	100%	100%	
	(514)*	(576)*	
	C. Education of Parent		
	Less Than High School	**High School**	**Some College**
0	75	60	56
1	17	27	26
2	8	13	18
	100%	100%	100%
	(335)*	(400)*	(345)*
	D. Race of Parent		
	White	**Black**	
0	59	79	
1	25	19	
2	16	2	
	100%	100%	
	(871)*	(151)*	
	E. Religiosity of Parent		
	Not Religious at All	**Moderate**	**Very Religious**
0	54	67	67
1	31	25	20
2	15	8	14
	100%	100%	101%
	(309)*	(239)*	(486)*

NOTE: Overall difference in each panel is significant at the .001 level. The source is merged data from the parent interviews in the 1981 National Survey of Children and from the 1983 grandparent survey. Weighted *n*'s are based on 1981 sampling weights.

* The number in parentheses represents a weighted *n*.

negatively. Panel *B* shows that interviews were obtained in 45 percent of the families in which the parent reported receiving any help from her parents or parents-in-law "such as childcare, errands, housework, or home repairs" in the past few weeks, versus 28 percent in families where no help was received. To be sure, we do not know whether the grandparent who was interviewed was the same one who lived close by or provided help, but it seems likely that nearby and helping grandparents were overrepresented. Similarly, panels *C, D,* and *E* suggest overrepresentation of grandparents whose children are better educated, less religious, and white.

How serious are these deviations from random representation? To illustrate their potential effects on our findings, let us perform the following thought experiment. In chapter 3, we classify grandparents according to their style of interaction with the grandchildren in our study, concluding that 29 percent are "remote," 55 percent are "companionate," and 16 percent are "involved." We note later in the text that the involved style is more common among grandparents who live nearby; and panel *A* suggests that nearby grandparents are overrepresented in the sample. By how much are we likely to have overestimated the proportion with involved relationships and underestimated the proportion with remote relationships?

The best answer is as follows. Suppose parents who reported that all of their parents lived an hour or more away had provided us with grandparent interviews at the same rates as did parents with nearby older parents. Applying the percentages in column two of panel *A* to the number of grandparents in each cell of column one of panel *A,* we estimate that 15 percent more grandparents would have been added to our sample, all of whom would have lived relatively far from the study children. Now consider that the grandparents in 1983 were asked how far they lived from the study children: "within 1 mile," "1 to 10 miles," "11 to 100 miles," or "more than 100 miles." (See question 1.) Let us make the rough but serviceable assumption that living within ten miles is equivalent to living within an hour's drive. Then, if you are still with us, consider the following table drawn from our 1983 data (and based upon the 1983 sampling weights we use throughout the text):

	Distance		
Relationship	**Within an Hour's Drive**	**Farther**	**Total**
Remote	3	51	29
Companionate	70	41	55
Involved	26	7	16
	99%	99%	100%
(Weighted *n*)	(324)	(368)	(692)

About half of all the grandparents who lived more than an hour's drive away (by our assumption about distance) had a remote relationship. Now let us add the additional 15 percent more grandparents to the table, all of whom would fall in the "farther" column. And let us make the rather pessimistic assumption that fully three-fourths of the additional grandparents had a remote relationship, one-fourth had a companionate relationship, and none had an involved relationship. Calculations then reveal that the expected distribution of the augmented—and presumably less biased—sample would be 35 percent remote, 52 percent companionate, and 14 percent involved.

In other words, we are likely to have underestimated the percent remote by about six percentage points and to have overestimated the percent companionate by three percentage points and the percent involved by two percentage points. These are modest differences; the use of these "corrected" figures would not cause us to alter our discussion in chapter 3. Specifically, our argument still stands that the companionate relationship is most common and the involved relationship least common. Our conclusion is that the selection bias in our sample is not large enough to distort the findings presented in the text. Nevertheless, our findings do present a slightly more positive picture of intergenerational support because of the underrepresentation of those who live farther away and who help out less. We have tried to be aware of this bias when interpreting findings in the text.

Our investigations of the 1981 parent data also reveal that, overall, maritally disrupted families are represented adequately. But among all grandparents in our sample whose children's marriages have disrupted, there clearly is a bias toward the maternal, custodial side. We were able to interview 122 grandparents whose children had custody after the divorce; we believe this to be a fairly representative sample. But we could obtain only thirty-two interviews with grandparents whose children did not have custody of their offspring following a marital disruption. The reason for this difference is that most of the former group are maternal grandparents, and most of the parents who were interviewed in 1981 were mothers. Clearly, divorced and separated mothers were more willing to give us the names of their parents than of their ex-in-laws. In chapter 6, we discuss evidence that they were especially reluctant to provide the names of ex-in-laws with whom they no longer had contact. Thus, our results are biased toward understating the differences between grandparents on the custodial side and those on the noncustodial side after a divorce in the middle generation. But as we note in chapter 6, in this case the bias strengthens our argument, for we find strong differences between the two groups of grandparents despite the probable underrepresentation of estranged grandparents on the noncustodial side.

We were more successful in obtaining interviews with maternal grandparents than with paternal grandparents in nondisrupted families for the same reason: it was the mothers who provided us with nearly all of the names. Even in families with no marital disruption, we obtained interviews with 200 maternal grandparents but only 145 paternal grandparents. We suspect that wives were more reluctant to give us access to their husband's parents—and probably even more reluctant if the husband's parents were infirm or they did not get along with their daughter-in-law. Fortunately, there is some information in the 1981 parent interview which we cite on this topic in chapter 6.

As chapter 1 notes, we were able to obtain just sixty-four interviews with grandfathers instead of the one hundred we had planned. The 446 grandmothers in our sample appear to be a representative group: the distribution of their educational attainments, for example, closely matches that of an age-standardized national population of middle-aged and older American women. But the sixty-four grandfathers are better educated than national figures would lead us to expect, suggesting that better-educated (and therefore probably younger) grandfathers were more willing to participate in the study. However, in our qualitative follow-up interviews we collected supplementary information about the differences between grandfathers and grandmothers, and we were able to combine the qualitative and quantitative information in a section of gender differences in chapter 5.

Table A–1 also demonstrated that black families were less likely to produce an interview than were white families. We interviewed fifty-one black grandparents, 10 percent of our unweighted sample; but when we weighted the sample to account for the intentional oversampling of maritally disrupted families, the proportion of blacks was reduced to 5 percent. Once again, we made a special effort to collect supplementary information in our qualitative interviews. Moreover, at several points in the text we give particular attention to patterns of grandparenting among blacks, most extensively in chapter 5.

There is one additional complication. Because we asked the parents for the names of grandparents on all sides, our 510 grandparents came from referrals by 380 parents. Thus, we have interviews with one grandparent for 254 study children, with two grandparents for 122 study children, and with three grandparents for 4 study children. Statistically speaking, then, we cannot consider the respondents to be fully independent observations. To check the potential bias of this departure from independence, we selected one grandparent at random from families that had produced more than one interview and reran several of the statistical analyses with this reduced sample. The results were quite similar. Nevertheless, we have interpreted the tests of statistical significance presented in the various appendices as suggestive rather than precise. And we have tried to focus our discussions on effects that seemed substantial in magnitude and that were borne out in our follow-up interviews.

Overall, this investigation of the representativeness of the sample suggests that there are some biases but that their effects are modest. Where more substantial biases do exist—as in the analysis of paternal versus maternal grandparents in maritally disrupted families—estimates can be made of the likely effect on the findings. We are confident that the major findings presented in this book are not distorted by sampling limitations. Still, we have been careful, throughout the book, to note the points at which potential biases most impinge upon our findings.

Despite the limitations, we believe this to be one of the better data sets on intergenerational relations that have been collected. Its national representativeness (with the caveats noted previously), its breadth of substantive coverage, and its relatively large size make it unique. Inevitably, any project that attempts to collect information from three generations in a family will have some deviations from random sampling. The investigator must start with one generation—in this case, a random sample of children. Had this project started in 1976 with a random sample of grandparents, and if the children's names had been referred by them, then the data for children would show some of the same limitations. In order to match data across three generations in the same families—as we have done in chapter 7—some compromises with random sampling must be made. Moreover, some of the same problems of overrepresentation—such as of the better educated or of whites—occur in standard surveys; it is unlikely that a direct sampling of grandparents would have been free of bias. Whatever our exact methodology, middle-class grandparents and those with more satisfactory relations would have been somewhat more likely to participate. Within the unavoidable constraints of survey research, then, there is much useful information about intergenerational relations to be gleaned from our data.

The Questionnaire

On pages 227 through 237, we present the questionnaire that was used in the 510 telephone interviews. Throughout the text the reader will be referred to this questionnaire for the exact wording of questions and the composition of attitude and behavior scales. Questions with numbers that are circled were asked of all respondents; all others were asked of a subset of respondents, depending upon their responses to previous questions. To save space, we have omitted instructions to the interviewers, such as "If divorced, skip to question 53" or "circle all that apply."

GRANDPARENT STUDY

Case #: ___________

Respondent's Name: ______________________________________

Address: ______________________________________

____

____

Telephone Number: ( ) ______________________

Interviewer: ______________________ ID#: ___________

Time began: ________ A.M.

________ P.M.

Date Completed: ________________ Time ended: ________ A.M.

________ P.M.

(1.) About how far from you does (CHILD) live: (a) within 1 mile; (b) 1 to 10 miles; (c) 11 to 100 miles; or (d) more than 100 miles? (e) Child lives with respondent.

2. How long has (CHILD) lived with you?

3. How long ago did you see (CHILD)? Today? Or ________ (DAYS) or ________ (WEEKS) or ________ (MONTHS) or ________ (YEARS)?

4. Was it: (a) a special occasion; or (b) just a regular get together?

5. What was the occasion?

6. Did it take place: (a) in your home; (b) in (CHILD)'s parents' home, or (c) someplace else (specify)?

7. Who else was there?

8. Who made the plans: (a) you; (b) (CHILD)'s parents; (c) (CHILD); (d) someone else (specify); or (e) no one?

9. In the past 12 months, about how often have you seen (CHILD): (a) almost every day; (b) 2 or 3 times a week; (c) about once a week; (d) once or twice a month; (e) once every 2 or 3 months; or (f) less often?

10. When you see (CHILD), who *usually* suggests getting together: (a) you; (b) (CHILD)'s parent; (c) (CHILD); or (d) who (specify)? (e) it varies; (f) no one.

11. When you saw (CHILD) in the past 12 months, was a parent with (him/her): (a) always; (b) usually; (c) occasionally; or (d) never?

12. Over the past 12 months, how often did you get to talk with (CHILD) on the telephone: (a) almost every day; (b) 2 or 3 times a week; (c) about once a week; (d) once or twice a month; (e) once every 2 or 3 months; or (f) less often?

13. Over the past 12 months, how many days, if any, did (CHILD) stay overnight in your home?

14. Have you and (CHILD) ever lived together for as long as 3 months? (a) Yes; (b) No.

15. What was the longest time you lived together? ____________
(MONTHS)

____________.
(YEARS)

16. How old was (CHILD) then?

____
(AGE OR AGE RANGE)

17. What were the circumstances?

18. Do you ever find that you don't spend as much time with (CHILD) as you would like because of the following: (a) Your health prevents it? (b) (CHILD) lives far away? (c) The trip is too difficult? (d) You are too busy? (e) (CHILD)'s parents are too busy? (f) (CHILD) is too busy? (g) Either of (CHILD)'s parents don't want you to see (him/her) more?

19. Would you say you spend: (a) too much time with (CHILD); (b) about the right amount of time; (c) a little less time than you would like; or (d) a lot less time than you would like?

(20.) How old is (CHILD) now? (a) ________ ; (b) Don't know.
(AGE)

(21.) When is (his/her) birthday? (a) ____________ ________ ;
(MONTH) (DAY)

(b) Don't know.

(22.) What grade is (CHILD) now in in school? (a) ____________ ;
(GRADE)

(b) Don't know; (c) Not in school.

23. Over the past 12 months, have you attended any of (his/her) school functions such as a school play, a sporting event, or graduation? (a) Yes; (b) No.

(24.) Can you and (CHILD) exchange ideas or talk about things that really concern you: (a) extremely well; (b) quite well; (c) fairly well; or (d) not very well?

(25.) Is your relationship with (CHILD): (a) extremely close; (b) quite close; (c) fairly close; or (d) not very close?

26. Here is a list of activities that grandparents sometimes do with their grandchildren. Over the past 12 months, did you: (a) Watch T.V. with (CHILD)? (b) Go to church or synagogue with (CHILD)? (c) Joke or kid with (him/her)? (d) Take a day trip together? (e) Give (CHILD) money? (f) Discipline (him/her)? (g) Give (CHILD) advice? (h) Talk to (him/her) about when you were growing up? (i) Teach (him/her) a skill or game? (j) Discuss (his/her) problems? (k) Help settle a disagreement between (CHILD) and (his/her) parents?

27. Over the past 12 months, has (CHILD): (a) Asked for your help with something (s/he) was doing or making? (b) Run errands or done chores for you? (c) Bought or made you something?

28. Over the past 12 months, have you: (a) Asked (CHILD) for help with something you were doing or making? (b) Helped (CHILD) with (his/her) errands or chores? (c) Bought or made something for (CHILD)?

(29.) In what way, if any, are (CHILD)'s parents bringing (him/her) up differently from how you brought up your children?

(30.) Have you had differences of opinion with (CHILD)'s parents about: (a) How (s/he) dresses? (b) What (s/he) eats? (c) How late (s/he) stays out at night? (d) How much to discipline (him/her)? (e) The friends (s/he) spends time with?

31. How often have you felt like expressing a difference of opinion but didn't, about the way (CHILD)'s parents are bringing (him/her) up: (a) often; (b) sometimes; (c) hardly ever; or (d) never?

32. When you see (CHILD) do something you disapprove of, do you correct (him/her): (a) often; (b) sometimes; (c) hardly ever; or (d) never?

33. Do your children consult you before making an important decision about (CHILD): (a) often; (b) sometimes; (c) hardly ever; or (d) never?

34. During the past year, have you: (a) Been to a wedding, funeral or religious ceremony for a relative? (b) Been to a large family get together that was just a social affair? (c) Exchanged a picture, letter, or family heirloom with one of your relatives?

35. When you think about your family, who specifically do you include?

36. Does your family share news and information about one another: (a) a great deal; (b) sometimes; or (c) hardly ever?

37. When there is important family news, are you: (a) one of the first to hear it; (b) one of the last; or (c) somewhere in between?

38. Which family member, if anyone, do you usually share information with?

39. Does your family gossip or tell stories about one another: (a) a great deal; (b) sometimes; (c) hardly ever; or (d) never?

40. Which family member, if anyone, do you gossip with?

41. Whom, if anyone, in your family are you not on speaking terms with?

42. In your family: (a) Are there special family recipes or dishes? (b) Are there special places that family members go away to together? (c) Are there family jokes, common expressions, or songs? (d) Are there ritual or special events that bring the family together? (e) Is there a family tree or history of the family?

43. Do you happen to know either of your grandmothers' maiden names? (a) Yes; (b) No.

44. How about your great grandmothers' maiden names? (a) Yes; (b) No.

(45.) Are any members of your family named after you? (a) Yes; (b) No.

(46.) Did you help choose any of your grandchildren's names? (a) Yes; (b) No.

(47.) How many living children and stepchildren do you have?

(48.) How many of them are stepchildren?

(49.) Let's start with your oldest (step) child. Is that a son or a daughter?

(50.) What is (his/her) first name?

(51.) Is (s/he) married, separated, divorced, widowed, or has (s/he) never married?

52. Has (s/he) ever been divorced?

(53.) How many living children and stepchildren does (s/he) have?

54. How many of these are stepchildren? (REPEAT QQ. 49–54 FOR EACH ADDITIONAL STEPCHILD AND CHILD.)

(55.) Have we missed any of your grandchildren? (a) Yes; (b) No.

56. How many?

57. Who are the children's parents?

(58.) Just so I'm sure, how many of your grandchildren, if any, are stepgrandchildren?

59. And just to be sure, is (CHILD) your natural or your stepgrandchild? (a) Natural; (b) Stepgrandchild.

60. Which of your children is (CHILD)'s parent?

61. (Counting natural and stepgrandchildren,) how old is your oldest grandchild?

62. (Counting natural and stepgrandchildren,) how old is your youngest grandchild?

63. How many of your grandchildren, if any, live with you now?

64. When was the last time you saw any grandchild (who doesn't live with you)? Today? Or ________ (DAYS) or ________ (WEEKS) or ________ (MONTHS) or ________ (YEARS) ?

65. It's not unusual for grandparents to like some grandchildren a little more than others. Do you have a favorite grandchild? (a) Yes; (b) No.

66. Is that a boy or a girl?

67. How old is (s/he)?

68. Which one of your children is the parent of that child?

69. Why is (s/he) your favorite grandchild?

70. How many great-grandchildren, if any, do you have?

71. What difference, if any, is there between being a grandparent and a great-grandparent?

72. What adults is (CHILD) living with now? (Probe): What other adults? (a) Mother; (b) Father; (c) Stepmother; (d) Stepfather; (e) Grandparent(s); (f) Other relatives; (g) Nonrelatives; (h) On own; (i) In correctional institution; (j) In other institution (specify); (k) In another place (specify).

73. Is your relationship with (CHILD)'s (natural) mother: (a) extremely close; (b) quite close; (c) fairly close; or (d) not very close? (f) Mother deceased; (g) No relationship.

74. Is your relationship with (CHILD)'s (natural) father: (a) extremely close; (b) quite close; (c) fairly close; or (d) not very close? (f) Father deceased; (g) No relationship.

75. Is your relationship with (CHILD)'s stepmother: (a) extremely close; (b) quite close; (c) fairly close; or (d) not very close?

76. Is your relationship with (CHILD)'s stepfather: (a) extremely close; (b) quite close; (c) fairly close; or (d) not very close?

77. Here are some statements about family life today. Tell me whether you strongly agree, mostly agree, mostly disagree, or strongly disagree that: (a) People should not get married unless they are deeply in love; (b) Marriages are better when the husband works and the wife runs the home and cares for the children; (c) Living together before marriage makes a lot of sense; (d) It should be easy for unhappy couples to get a divorce; (e) Children are better off if their mothers do not work outside the home; (f) Single women should not have children, even if they want to; (g) When parents divorce, children develop permanent emotional problems; (h) There is no reason for married couples to have children if they prefer not to; (i) Unless a couple is

prepared to stay together for life, they should not get married; (j) After a divorce, the mother should automatically get custody of all children.

(78.) Is either of your parents living? (a) Yes; (b) No.

79. Does either of your parents live within an hour's drive of you? (a) Yes; (b) No; (c) Parent lives with respondent.

80. How often do you see either of your parents: (a) almost every day; (b) once or twice a week; (c) once or twice a month; (d) occasionally during the year; (e) hardly ever; or (f) never? (h) Other (specify).

81. In the past few weeks have you provided any help, such as errands, housework, or home repairs, to either of your parents? (a) Yes; (b) No.

82. Over the past 12 months, have you provided any financial support to them? (a) Yes; (b) No.

83. And have you received any financial help from them? (a) Yes; (b) No.

(84.) Did you know any of your grandparents when you were growing up? (a) Yes; (b) No.

85. Did any of them ever live in your house while you were growing up? (a) Yes; (b) No.

86. Are you and (CHILD): (a) more friendly; (b) less friendly; or (c) about the same as your grandparents were with you?

87. Do you have: (a) more authority over (CHILD); (b) less authority over (CHILD); or (c) about the same as your grandparents had over you?

88. Is your relationship with (CHILD): (a) closer; (b) not as close; or (c) about the same as your grandparents' was with you?

89. Were you: (a) more respectful; (b) less respectful; or (c) about the same toward your grandparents as (CHILD) is toward you?

(90.) How old was (CHILD) when you most enjoyed being (his/her) grandparent? ________________________.

(AGE OR AGE RANGE)

91. I'll read you a list of things that sometimes happen to grandparents when their children's marriages break up: (a) Did (CHILD)'s parents try to keep their marital troubles from you before the breakup? (b) Did (CHILD)'s parents tell you about their plans to separate before the breakup occurred? (c) Did you try to prevent the breakup from occur-

ring? (d) Did you discuss with either of (CHILD)'s parents what effects the breakup might have on (CHILD)? (e) Did you talk directly with (CHILD) about the breakup? (f) Did you provide some financial help to (CHILD) or either of (his/her) parents? (g) Did (CHILD) come to live with you around the time of the breakup?

92. During the breakup, did you see (CHILD): (a) more often; (b) less often; or (c) about the same amount of time as before their marital problems began?

93. How about now, do you see (CHILD): (a) more often; (b) less often; or (c) about the same amount of time as you did before the breakup?

94. Is your relationship with (CHILD): (a) as close; (b) not as close; or (c) closer than it was before the breakup?

95. Overall, how did the breakup change your relationship with (CHILD)?

96. What do your grandchildren call you?

97. What do your natural grandchildren call you?

98. What do your stepgrandchildren call you?

99. I'll read a list of statements about being a stepgrandparent. Based on your experience with your stepgrandchildren and your natural grandchildren, do you agree very much, somewhat, or not at all that: (a) It is generally harder for you to love your stepgrandchildren than to love your own grandchildren; (b) It is harder for you to be a stepgrandparent than a natural grandparent; (c) Your stepgrandchildren can't think of you as a real grandparent; (d) You've had problems accepting your stepgrandchildren as grandchildren; (e) You feel differently about your natural grandchildren and your stepgrandchildren; (f) You tend to see less of your stepgrandchildren; (g) Your stepgrandchildren seem less interested in you than your own grandchildren do.

100. After you die, will some portion of your possessions go directly to your grandchildren (or stepgrandchildren)? (a) Yes; (b) No; (c) Don't know.

101. How will your possessions be divided among your grandchildren (and stepgrandchildren)?

102. Are you currently: (a) married; (b) widowed; (c) separated; (d) divorced; or (e) have you never married?

103. How many times have you been married?

104. What is your religious preference, if any? (a) Protestant; (b) Roman Catholic; (c) Jewish; (d) Other (specify); (e) Atheist, agnostic, or none.

105. What denomination is that?

106. How important is religion to you? (a) Very important; (b) Fairly important; (c) Not very important.

107. Do you attend religious services or activities: (a) about once a week; (b) at least once a month; (c) a few times a year; or (d) never?

108. What is the highest grade of school you have completed? (a) No formal schooling; (b) Elementary school (01 02 03 04 05 06); (c) High school (07 08 09 10 11 12); (d) College or trade school (13 14 15 16); (e) Graduate or professional school (17 18 19 20+).

109. What diplomas or degrees, if any, did you receive?

110. Are you presently employed, unemployed, retired, keeping house, or what? (a) Employed; (b) With a job, but not at work because of temporary illness, sick leave, vacation, labor dispute, bad weather; (c) Retired; (d) Keeping house; (e) Disabled; (f) Other (specify); (g) Unemployed; (h) In the armed services.

111. Are you now looking for work? (a) Yes; (b) No.

112. In what month and year (if ever), did you last work for pay at a regular job or business, either full time or part time? (a) ____________ ____________ ; (b) Never.

(MONTH) (YEAR)

113. (Is/Was) that work full time or part time?

114. What (is/was) your main occupation? That is, what (is/was) the job title?

115. Tell me about what you actually (do/did) at the job. What (are/were) the main duties?

116. What kind of business or industry (is/was) that? (Probe: What do they do or make there?)

117. Would you say your own health, in general, is: (a) excellent; (b) good; (c) fair; or (d) poor?

118. How much do health troubles stand in the way of your doing the things you want to do? (a) Not at all; (b) A little; or (c) A lot.

119. Because of your health, do you need help from others in looking

after your personal needs such as eating, dressing, undressing, or personal hygiene? (a) Yes; (b) No.

120. Do you need help from others to go outdoors or to get around outside the house? (a) Yes; (b) No.

121. Within the last year, have you received any income from: (a) Social Security? (b) Unemployment insurance or compensation? (c) Food stamps? (d) Welfare or other public assistance?

122. From all sources of income (including those you have mentioned), was your total family income before taxes in 1982: (a) Under $5,000; (b) $5,000 to $10,000; (c) $10,000 to $15,000; (d) $15,000 to $20,000; (e) $20,000 to $25,000; (f) $25,000 to $35,000; (g) $35,000 to $50,000; or (h) $50,000 or over? (If uncertain): What would be your best guess?

123. How many people depend on this income?

124. Over the past 12 months, have you received any financial support from (CHILD)'s parents? (a) Yes; (b) No.

125. Over the past 12 months, have you provided any financial support to (CHILD)'s parents? (a) Yes; (b) No.

126. Over the past 12 months, did you provide financial support: (a) regularly; (b) occasionally; or (c) seldom?

127. Was this financial support: (a) a loan with pay-back expected; or (b) a gift with no strings attached?

128. What is your age?

129. What is your (husband's/wife's) age?

130. All things considered, is your life going: (a) very well; (b) fairly well; (c) not so well; or (d) not well at all?

131. How do you describe yourself in terms of your racial or ethnic background: (a) American Indian; (b) black or Afro-American; (c) Mexican-American or chicano; (d) Puerto Rican or other Latin American; (e) Oriental or Asian-American; (f) white or caucasian; or (g) some other ethnic group (specify)?

132. In appreciation for your time, the Institute would like to send you a report of the National Survey of Children in which (CHILD) participated. Please spell your name and tell me your mailing address.

INTERVIEWER'S OBSERVATION SECTION TO BE COMPLETED AT THE CONCLUSION OF THE INTERVIEW.

The Grandparent Study

1. Rate respondent's apparent intelligence: (a) Very high; (b) Above average; (c) Average; (d) Below average; (e) Very low.

2. Which questions, if any, did the respondent have difficulty understanding?

3. In general, how quickly did the respondent respond to questions? (a) Responded quickly, without hesitation; (b) Deliberated some, but responses were generally not too slow; (c) Was often slow to respond; (d) Was usually very slow to respond, needed much urging.

4. During the interviewing, was the respondent generally: (a) very interested or enthusiastic; (b) somewhat interested; (c) indifferent; (d) somewhat reluctant; or (e) very reluctant?

5. How attentive was the respondent during the interview? (a) Attentive, involved, responsive; (b) Somewhat inattentive and uninvolved; (c) Easily distracted, needed urging to pay attention.

6. How truthful did the respondent seem to be? (a) Completely truthful; (b) Usually truthful; (c) Often untruthful; (d) Mainly untruthful, evasive, or incoherent.

7. At the end of the interview, did the respondent seem to be: (a) fairly tired; (b) a little tired; or (c) not tired at all?

8. What else is there about the interview that will help in interpreting the data?

Appendix 2

Statistical Analyses for Chapter 5

Frequency of Contact

The response categories to question 9, "In the past twelve months how often have you seen the child," were transformed to indicate the annual number of visits. (The untransformed response frequencies are shown in the text in chapter 3.) Grandparents who were living in the same household as the study child (4 percent of our sample) were excluded from the analysis of frequency of visits. For some response categories, such as "about once a week," the appropriate transformation was clear—in this case, 52. But for response categories such as "once or twice a month" the appropriate transformation was unclear. We tried several plausible ways of transforming the response categories, and the results of the analyses were virtually identical. The transformation we report on is as follows: "almost every day": (5 × 52 =) 260; "two or three times a week": (2.5 × 52 =) 130; "about once a week": (1 × 52 =) 52; "once or twice a month": (1.5 × 12 =) 18; "once every two or three months": (12 ÷ 2.5 =) 4.8; and "less often": 1. (Grandparents who had not seen the study child in the past twelve months also were given a transformed score of one to allow for a further logarithmic transformation to be described.)

This transformation provided us with a measure of variation in the absolute frequency of contact, but we believe that it is the proportional variation in contact that is more meaningful. To give an example, we believe that the difference between seeing a grandchild once a month and once a week is as about as important substantively as between seeing a grandchild once a week and almost every day. In order to better reflect this type of proportional variation, we used the natural logarithm of the annual number of visits as our dependent variable. The logarithmic transformation is used frequently in linear models for which proportional change in the dependent variable is to be explained. The natural logarithms of the six transformed values are, respectively, 5.56, 4.87, 3.95, 2.89, 1.57, and .69. These are the values our dependent variables take in the subsequent analyses.

We then regressed the natural logarithm of the annual frequency of visits on a set of independent variables according to the following equation:

$$ln(y) = a + b_1x_1 + b_2x_2 + \cdots + b_nx_n,$$

where y is the annual number of visits, the x_i are sets of dummy variables coded zero or one, and the b_i are parameters to be estimated. Ordinary least squares estimation was used. In this type of semilogarithmic specification, it is sometimes easier to interpret the parameters in the multiplicative form of the equation, which is obtained by raising both sides to the base e:

$$y = exp(a)exp(b_1x_1)exp(b_2x_2) \cdots exp(b_nx_n).$$

Each multiplicative coefficient $exp(b_i)$ shows the estimated proportional change in y for a unit change in x_i. If the x_i are dummy variables, as in our analysis, then the $exp(b_i)$ show the proportional change in y relative to the level of y for the omitted category. For example, suppose $x_1 = 1$ for females and $x_1 = 0$ for males. Then, other variables held constant, the predicted value of y is multiplied by $exp(b_1 \times 1) = exp(b_1)$ for females and is multiplied by $exp(b_1 \times 0) = 1$ for males. Thus, the predicted value of y for females is $exp(b_1) \div 1 = exp(b_1)$ times the value for males. In other words, females see their grandchildren, other variables held constant, $exp(b_1)$ times more (or less) frequently per year than do males. We will present these multiplicative coefficients, which can be interpreted as the predicted increase (if greater than one) or decrease (if less than one) in annual number of visits relative to the omitted category in each set of dummy variables. A multiplicative coefficient of one implies no difference relative to the omitted category.

The second dependent variable was formed in exactly the same manner from the question, "Over the past twelve months, how often did you get to talk to [the child] on the telephone?" Since this question was not asked of those who lived with the study child or who had not seen the study child in the past year (a total of 13 percent of the sample), these grandparents were excluded.

The independent variables are listed in table A–2. All take the form of dummy variables coded one if true and zero if false. Distance information comes from the responses to question 1. (See appendix 1 for a reproduction of the questionnaire.) The information on the relationship to the study child's mother was obtained from question 73 for most children; but for the small number of children who were not living with their mothers, the responses to question 74 (relationship with father) were substituted. (Cases in which the study children did not live with either parent were dropped from this analysis.) Information on financial support was obtained from question 125. (We also asked about the frequency of financial support in question 126,

Statistical Analyses for Chapter 5

TABLE A–2

*Independent Variables with Weighted Frequencies for Analysis of Frequency of Visiting**

Variables	Frequency Distribution
Distance: within 1 mile	15
Distance: 1 to 10 miles	31
Distance: 11 to 100 miles	22
Distance: more than 100 miles†	32
	100%
Relationship to mother: extremely close	48
Relationship to mother: quite close	30
Relationship to mother: fairly close or not very close†	22
	100%
Family rituals: "special family recipes or dishes" *and* "family jokes, common expressions, or songs"	54
Family rituals: one or none of above†	46
	100%
Lineage: maternal	57
Lineage: paternal†	43
	100%
Provided financial support: yes	16
Provided financial support: no†	84
	100%
Size of place: less than 2,500	35
Size of place: 2,500–99,999	46
Size of place: 100,000 or more	19
	100%
Weighted *n*	666

* All variables are coded one if true and zero if false.
† Omitted category in the regression analyses.

but further analyses showed that, among those who provided support, variations in frequency were unrelated to contact.) The family ritual variable was coded one if the respondent responded affirmatively to both question 42a and question 42c;

TABLE A–3

Regression of Natural Logarithm of Annual Frequency of Visits with Grandchild on Selected Indicators

Variables	Multiplicative Coefficients (1)	Multiplicative Coefficients (2)
Distance: within 1 mile	38.5*	33.1*
Distance: 1 to 10 miles	15.2*	13.6*
Distance: 11 to 100 miles	4.83*	4.04*
Relationship to mother: extremely close	—	2.04*
Relationship to mother: quite close	—	1.58*
Family rituals: both	—	1.46*
Lineage: maternal		1.16
Provided financial support: yes	—	1.32†
Size of place: less than 2,500	—	1.28†
Size of place: 2,500–99,999	—	0.83
Multiplicative constant	2.65	1.29
Proportion of variance accounted for (R^2)	0.625	.688
Weighted *n*	666	666

NOTE: Grandparents who were living with their grandchildren were excluded from the analysis.
* Additive coefficient is more than three times its standard error.
† Additive coefficient is more than twice its standard error.

otherwise it was coded zero. The lineage variable was coded one ("maternal") if the respondent was related to the study child through a daughter and zero if related through a son ("paternal"). And the size-of-places variable refers to the population size of the place of residence of the grandparent.

Results

Table A–3 displays the results of the regression analyses of frequency of visits, which are discussed in chapter 5. First, frequency of visits was regressed only on the set of dummy variables indicating distance. These results are shown in column one. Then frequency of visits was regressed on the complete set of independent variables, with the results shown in column two. Table A–4 displays an analogous regression of the natural logarithm of the annual frequency of telephone calls on the same set of variables as in column two of table A–3.

Parentlike Behavior and Exchanges of Services

As mentioned in the text, the answers to two sets of questions about activities tended to cluster together. This clustering means that, within each set, knowing how a person answered one question allows the analyst to predict with some degree of

TABLE A–4

Regression of Natural Logarithm of Annual Frequency of Telephone Conversations with Grandchild on Selected Indicators

Variables	**Multiplicative Coefficients**
Distance: within 1 mile	4.05*
Distance: 1 to 10 miles	4.33*
Distance: 11 to 100 miles	1.79*
Relationship to mother: extremely close	1.62†
Relationship to mother: quite close	1.67†
Family rituals: both	1.09
Lineage: maternal	1.56*
Provided financial support: yes	1.32
Size of place: less than 2,500	0.61*
Size of place: 2,500–99,999	0.60*
Multiplicative constant	4.62
Proportion of variance accounted for (R^2)	0.202
Weighted *n*	603

NOTE: Grandparents who were living with the study child or who had not seen the study child within the previous 12 months were excluded from the analysis.
* Additive coefficient is more than three times its standard error.
† Additive coefficient is more than twice its standard error.

accuracy how that person responded to the other questions. In other words, each set taps a related group of activities. We identified these sets using the statistical technique of factor analysis (specifically, a principal factor solution followed by a varimax rotation), and then formed scales that indicate the extent to which each grandparent engages in each set of activities. The first scale we refer to as "parentlike behavior." This scale measures the extent to which a grandparent has the authority to play an active role in raising the child. It consists of responses to questions 26f, 26g, 26j, 32, and 33. The meaning of this scale will be clearer if the reader inspects these five questions in the questionnaire in appendix 1. We computed scores on the scale of parentlike behavior by summing, for each grandparent, the number of positive responses to 26f, 26g, and 26j and the number of "often" or "sometimes" responses to 32 and 33. The result was a scale score that ranged from zero to five, with a score of five representing the greatest amount of parentlike behavior. (The alpha reliability of the scale was 0.74.) The second scale measures the exchange of services between grandparents and grandchildren, and it was formed the same way. It consists of responses to questions 27a, 27b, 28a, and 28b. A scale score that ranged from zero to four was computed by summing the number of positive responses to these questions. (The alpha reliability of the scale was 0.73.)

As noted in the text, scores on these two scales were used to classify a grandparent as "involved" or "companionate." Therefore, in our analysis of which grandparents were more likely to be involved, we examined the determinants of high or low scores on these scales. Specifically, we estimated a series of ordinary least-squares regression models with the parentlike behavior scale as the dependent variable and a second, identical series with the exchange scale as the dependent variable. In both series the independent variables were the same: frequency of contact with the grandchild in the past twelve months, age of grandparent, family ritual scale, race of grandparent, total family income of grandparent in 1982, grandparent's relationship with the study child's mother, and whether the study child's parents were separated or divorced. For these regressions, the family ritual scale was coded as the sum of the number of positive responses to questions 42a and 42c. (The answers to these two questions clustered together, according to our factor analyses.) As in the first section of this appendix, the relationship with mother question was taken directly from the responses to question 73 for most children; but for the small number of children who were not living with their natural mothers, the responses to question 74 were substituted. (Again, cases in which the study child did not live with either parent were dropped from this analysis.)

The results are presented in tables A–5 and A–6. Since all independent variables were coded in "dummy variable" form, we have followed a method of presentation similar to what is known as "multiple classification analysis." (See Frank M. Andrews, James N. Morgan, and John A. Sonquist, *Multiple Classification Analysis* [Ann Arbor: Institute for Social Research, 1967].) Column one shows the "unadjusted" mean scores on the dependent variable for each category of an independent variable. These unadjusted means represent the effect of the independent variable considered by itself. The "adjusted means" in the second column show the effect of an independent variable controlling for all other variables in the model. The proportion of variance accounted for (R^2) by each variable is presented for both the unadjusted and adjusted means. The former R^2 represents the predictive power of a variable considered by itself—that is, when it is the only independent variable in the regression equation. This R^2 comprises both the unique predictive power of a variable and the predictive power it shares with other variables. The latter R^2 represents the increment in predictive power obtained by entering that independent variable into the regression equation after all other variables have been entered—in other words, its unique predictive power alone. Tests of significance of both R^2 values were calculated.

TABLE A–5

Unadjusted and Adjusted Mean Scores on a Scale of Parentlike Behavior (Grand Mean: 2.27)

	(1) Unadjusted Means	(2) Adjusted Means	*N*
Frequency of contact with study child in past 12 months			
Almost every day	3.96	3.67	74
2 or 3 times a week	2.77	2.61	74
About once a week	2.53	2.45	98
Once or twice a month	1.92	2.00	125
Less often	1.56	1.72	201
	(0.223)*	(.118)*	
Age of grandparent			
45–59	2.93	2.66	115
60–64	2.69	2.71	113
65–69	2.02	1.99	151
70–87	1.83	2.00	193
	(.073)*	(.037)*	
Family rituals			
2	2.60	2.53	336
1	1.91	1.94	161
0	1.58	1.82	75
	(.058)*	(.004)	
Race			
Black	3.83	3.37	25
Nonblack	2.20	2.22	547
	(.040)*	(.018)*	
Family income in 1982			
Under $5,000	2.34	2.44	105
$5,000 to $10,000	2.05	2.13	174
$10,000 to $15,000	2.58	2.42	122
$15,000 to $20,000	1.69	1.95	66
$20,000 or more	2.57	2.36	105
	(.033)*	(.010)	
Grandparent's relationship with study child's mother			
Extremely close	2.46	2.28	289
Quite close	2.23	2.40	180
Fairly close	1.87	2.10	78
Not very close	1.64	2.04	25
	(.021)†	(.004)	
Study child's parents are separated or divorced			
Yes	2.71	2.64	124
No	2.15	2.17	448
	(.019)*	(.013)*	

NOTE: Numbers in parentheses are increments in R^2. The R^2 for the complete equation is .342.

* Significant at the 1 percent level.

† Significant at the 5 percent level.

TABLE A–6

Unadjusted and Adjusted Mean Scores on a Scale of Exchange of Services Between Grandparents and Grandchildren (Grand mean: 1.81)

	(1) Unadjusted Means	(2) Adjusted Means	*N*
Frequency of contact with study child in past 12 months			
Almost every day	3.42	3.27	75
2 or 3 times a week	2.60	2.47	74
About once a week	2.00	1.92	99
Once or twice a month	1.45	1.50	127
Less often	1.05	1.15	200
	(.301)*	(.190)*	
Age of grandparent			
45–59	2.38	2.14	113
60–64	1.87	1.87	113
65–69	1.87	1.82	151
70–87	1.39	1.57	197
	(.058)*	(.016)*	
Family rituals			
2	2.08	1.95	336
1	1.63	1.70	161
0	0.99	1.44	77
	(.064)*	(.013)*	
Race			
Black	1.90	1.57	29
Nonblack	1.80	1.82	546
	(.00)	(.001)	
Family income in 1982			
Under $5,000	1.58	1.74	106
$5,000 to $10,000	1.63	1.74	177
$10,000 to $15,000	2.12	1.88	123
$15,000 to $20,000	1.54	1.66	65
$20,000 or more	2.14	1.98	105
	(.031)*	(.005)	
Grandparent's relationship with study child's mother			
Extremely close	2.03	1.84	291
Quite close	1.81	1.96	179
Fairly close	1.31	1.53	79
Not very close	0.77	1.28	25
	(.048)*	(.013)†	
Study child's parents are separated or divorced			
Yes	1.84	1.81	122
No	1.80	1.79	452
	(.00)	(.00)	

NOTE: Numbers in parentheses are increments in R^2. The R^2 for the complete equation equals .342.

* Significant at the 1 percent level.

† Significant at the 5 percent level.

Appendix 3

Statistical Analyses for Chapter 6

In table A–7 we present crosstabulations of five indicators of the grandparent-grandchild relationship by grandparental status. The indicators are geographical distance (question 1), frequency of contact (question 9), provision of financial support (questions 125 and 126), and two scales discussed in detail in appendix 2: parentlike behavior and exchange of services. Grandparental status has four categories: (1) maternal grandparent with an adult daughter still in an intact first marriage; (2) maternal grandparent with an adult daughter who has ever separated or divorced; (3) paternal grandparent with an adult son still in an intact first marriage; and (4) paternal grandparent with an adult son who has ever separated or divorced. (The adult daughters and sons are the parents of the study children.)

We have omitted two smaller groups of grandparents from these crosstabulations. Because our focus is on marital separation and divorce, we eliminated grandparents of study children who were not living with both parents for reasons other than marital dissolution. Many of these study children were born out of wedlock. Their grandparents constituted 6 percent of the sample. Second, we eliminated "stepgrandparents," the older parents of the study children's stepparents. These stepgrandparents were of interest to us, but we were only able to interview twelve cases that were the result of a divorce followed by a remarriage. This number was too small to support the detailed tabulations reported in table A–7.

TABLE A–7

Comparisons of Grandparents in Nondisrupted and Disrupted Families, by Lineage

	Maternal Grandparents		Paternal Grandparents	
	Nondis-rupted	Dis-rupted	Nondis-rupted	Dis-rupted
A. About how far away from you does the child live?				
Child lives with respondent	1	13	1	2
Within 1 mile	14	11	18	1
1 to 10 miles	27	23	36	39
11 to 100 miles	24	22	15	20
More than 100 miles	34	31	30	37
	100%	100%	100%	100%
Weighted *n*	277	100	222	42
Chi-squared = 58.6, $p < .001$				
B. In the past 12 months, how often have you seen the child?				
Almost every day	8	23	15	4
2 or 3 times a week	14	13	11	1
About once a week	16	12	16	17
Once or twice a month	21	14	18	22
Once every 2 or 3 months	15	8	12	21
Less often	26	30	29	35
	100%	100%	101%	100%
Weighted *n*	277	100	222	42
Chi-squared = 33.2, $p < .01$				
C. Scale of parentlike behavior toward the study child.				
0–1 (low)	40	17	43	49
2–3	40	26	31	25
4–5 (high)	21	57	26	26
	101%	100%	100%	100%
Weighted *n*	247	91	203	38
Chi-squared = 48.4, $p < .001$				
D. Scale of exchange of services with the study child.				
0 (low)	27	17	32	48
1–2	36	33	35	28
3–4 (high)	37	59	33	25
	100%	100%	100%	100%
Weighted *n*	248	90	204	38
Chi-squared = 17.4, $p < .01$				
E. Over the past 12 months, have you provided any financial support to the child's parents? [If yes:] Over the past 12 months, did you provide financial support: regularly, occasionally, or seldom?				
Regularly	0	11	4	2
Occasionally	6	8	6	9
Seldom	4	13	6	7
No support provided	90	68	84	82
	100%	100%	100%	100%
Weighted *n*	277	100	222	42
Chi-squared = 40.6, $p < .001$				

NOTE: Percentages may not add to 100 due to rounding error.

Appendix 4

Statistical Analyses for Chapter 7

In table A–8 we display the nine common statements about attitudes toward family life that were included in each of the grandparent, parent, and study child (grandchild) surveys. The percent from each sample who agree with the item is shown separately for the three generations. For each of the grandparents in the survey, the responses of the associated grandchildren and parents are included. Since, in a minority of families, more than one grandparent was interviewed, the replies of some parents and children are counted more than once. (See appendix 1.) Children and parents in the 1981 interview were excluded if no grandparent was interviewed in 1983.

In table A–9 we show the correlations measuring intergenerational agreement for pairs of generations on both the traditional and modern scales of family values. The traditional scale is based on the responses to the following three items:

Q77b Marriages are better when the husband works and the wife runs the home and cares for the children.

TABLE A–8

Attitudes toward Family Life among Grandparents, Parents, and Grandchildren in the National Survey of Children

Question	Percent Agreeing:		
	Grandparent[a]	**Parent[b]**	**Grandchild[b]**
People should not get married unless they are deeply in love	91	93	95
Marriages are better when the husband works and the wife runs the home and cares for the children	76	58	55
Living together before marriage makes a lot of sense	6	23	42
It should be easy for unhappy couples to get a divorce	38	42	56
Children are better off if their mothers do not work outside the home	69	61	59
When parents divorce, children develop permanent emotional problems	69	48	68
There is no reason for married couples to have children if they prefer not to	80	94	90
Unless a couple is prepared to stay together for life, they should not get married	76	74	71
After a divorce, the mother should automatically get custody of all children	27	19	35

[a] Responses of grandparents to questions 77a through 77e and 77g through 77j in the 1983 survey. Percentage agreeing is the percentage responding either "strongly agree" or "mostly agree."

[b] Responses to the same questions in the 1981 survey of children and their parents.

Q77e Children are better off if their mothers do not work outside the home.

Q77g When parents divorce, children develop permanent emotional problems.

The contemporary family value scale is constructed from responses to the following three items:

Q77c Living together before marriage makes a lot of sense.

Q77d It should be easy for unhappy couples to get a divorce.

Q77i Unless a couple is prepared to stay together for life, they should not get married.

TABLE A–9

Correlations of Intergenerational Agreement on Traditional and Modern Scales

	Generation Pair		
	Grandparent/ Parent	**Grandparent/ Grandchild**	**Parent/ Grandchild**
Simple Correlations			
Scales			
Traditional	.225*	.071*	.334*
	(680)†	(677)	(692)
Modern	.209*	.077*	.330*
	(680)	(679)	(697)
Partial Correlations Controlling for Parent Influence			
Scales			
Traditional		−.004	
		(673)	
Modern		.006	
		(676)	

* Significant at .05 level.

† Weighted sample sizes are in parentheses.

Possible responses to each item included "strongly disagree," "disagree," "depends," "agree," and "strongly agree." A value from zero to four was assigned to each item based on the response such that the higher score reflected either a more modern response if the item was from the contemporary scale or a more traditional response if the item was from the traditional scale. A score for each scale was then derived by simply adding the scores of the three respective items. Therefore, the range of each scale varied from zero, indicating that the respondent was very low on either the modern or traditional dimension, to twelve, indicating a highly modern or traditional attitude. Modern and traditional scales were identically constructed for each generation—grandparents, parents, and grandchildren.

Intergenerational agreement on these scales was measured for three pairs of generations: grandparent and parent, grandparent and grandchild, and parent and grandchild. The simple correlations reveal, then, the strength of the association between each pair of generations on the traditional dimension and on the modern dimension. The sample size, again, is restricted to the grandparents in our survey and the associated grandchildren and parents. Some attrition in the analyses occurred due to nonresponses or "don't know" responses, and the sample size for each correlation is noted in parentheses.

Partial correlation analysis was also run to control for the influence of the parent with regard to the agreement of family values between the grandparent and the grandchild. With parental attitudes held constant, the partial correlations for the grandparent and grandchild on both the traditional and modern scales are shown in the lower half of table A–9. The residual correlation reflects the potential direct influence of the grandparents on the children, which is negligible.

Finally, in the last series of analyses reported in chapter 7, we use eight scales measuring the child's behavior. The procedures for developing these scales are discussed in Frank F. Furstenberg, Jr., and Paul A. Allison, "How Divorce Affects Children:

Variations by Age and Sex," a paper presented at the annual meeting of the Society for Research on Child Development, Toronto, April 1985.

Briefly, the items were selected from a large pool of information obtained from parents, teachers, and children in the 1981 National Survey of Children. The scales were developed by a series of screening procedures that put together items that had both high validity and reliability. Each subpool of items was further refined by testing it for unidimensionality, using confirmatory factor analysis. The final step was to construct scores for each scale by summing the number of positive responses. The scales used in the analyses reported in chapter 7 were as follows, with alpha reliabilities in parentheses:

From Parent's Report:

1. *Delinquency* (alpha = .60)
    1. Since January 1977, about the time of the first interview, has (he/she) had any behavior or discipline problems at school resulting in your receiving a note or being asked to come in and talk with the teacher or principal?
    2. Has (child) been suspended, excluded, or expelled from school since January 1977?
    3. Since January 1977, has (he/she) run away from home?
    4. Since January 1977, has (child) stolen anything, regardless of its value?
    5. How many times, if any, has (child) been stopped or questioned by the police or juvenile officers?
2. *Problem Behavior* (alpha = .69)

    Tell me whether each (of the following) statement(s) has been . . . true of (child) during the past three months:
    1. Cheats or tells lies.
    2. Is disobedient at home.

3. Is disobedient at school.
4. Hangs around with kids who get into trouble.

3. *Distress* (alpha = .69)
   Tell me whether each (of the following) statement(s) has been . . . true of (child) during the past three months:
   1. Has sudden changes in mood or feelings.
   2. Feels or complains that no one loves (him/her).
   3. Is too fearful or anxious.
   4. Feels worthless or inferior.
   5. Is unhappy, sad, or distressed.

From Teacher's Report:

4. *Problem Behavior* (alpha = .79)
   1. In your class, how often was any disciplinary action required for this student?

   For each of the following statements please indicate . . . how much like that this student was in 1980–81:
   2. Fought too much, teased, picked on, or bullied other students.
   3. Cheated, told lies, was deceitful.
   4. Had a very strong temper, lost it easily.

5. *Academic Difficulty* (alpha = .95)
   How did this student compare with others in (his/her) class last year (1980–81)?
   1. Verbal ability?
   2. Math ability?
   3. Overall performance?

From Child's Report:

6. *Delinquency* (alpha = .52)
   1. How many times, if ever, have you been stopped or questioned by the police or juvenile officers about something they thought you did wrong?

   In the last year, about how many times have you:

2. Hurt someone badly enough to need bandages or a doctor?
3. Lied to your parent(s) about something important?
4. Taken something from a store without paying for it?
5. Damaged school property on purpose?

7. *Dissatisfaction* (alpha = .71)
   Are you satisfied, somewhat satisfied, or not too satisfied with:
   1. Your friends?
   2. Your family?
   3. Yourself?
   4. Being a (boy/girl)?
   5. Being an American?

8. *Distress* (alpha = .46)
   1. Do you feel lonely and wish you had more friends?
   2. Do you have days when you are nervous, tense, or on edge?
   3. Do you have days when you are unhappy, sad, or depressed?
   4. All things considered, (how) is your life going?

NOTES

Chapter 1 / Introduction

1. Karl Abraham, "Some Remarks on the Role of Grandparents in the Psychology of Neurosis," in *Selected Papers and Essays in Psychoanalysis,* vol. 2, ed. Hilda C. Abraham (New York: Basic Books, 1955); and E. A. Rappaport, "The Grandparent Syndrome," *Psychoanalytic Quarterly* 27 (1958): 518–38.
2. See Ernest W. Burgess and Harvey J. Locke, *The Family: From Institution to Companionship* (New York: American Book Company, 1945); and Talcott Parsons and Robert F. Bales, *Family, Socialization, and the Interaction Process* (New York: The Free Press, 1955). Neither of these classic works contains an index entry for grandparents or intergenerational relations.
3. See Lillian E. Troll, "Grandparents: The Family Watchdogs," in *Family Relationships in Later Life,* ed. T. Brubaker (Beverly Hills, CA: Sage Publications, 1983), 63–74.
4. Arthur Kornhaber and Kenneth L. Woodward, *Grandparents/Grandchildren: The Vital Connection* (New York: Doubleday, Anchor Press, 1982).
5. The results are described in Nicholas Zill, *Happy, Healthy, and Insecure: A Portrait of Middle Childhood in America* (New York: Cambridge University Press, forthcoming).

Chapter 2 / The Modernization of Grandparenthood

1. Peter Uhlenberg, "Demographic Change and the Problems of the Aged," in *Aging from Birth to Death,* ed. Matilda White Riley (Boulder, CO: Westview Press, 1979), 153–66; and Uhlenberg, "Death and the Family," *Journal of Family History* 5 (Fall 1980): 313–20.
2. U.S. Bureau of the Census, *Historical Statistics of the United States: Colonial Times to 1970* (Washington, D.C.: U.S. Government Printing Office, 1975), ser. B120 and B121; and U.S. Bureau of the Census, *Statistical Abstract of the United States: 1984* (Washington, D.C.: U.S. Government Printing Office, 1985), 73 and 33.
3. Uhlenberg, "Death and the Family."
4. Lillian E. Troll, Sheila J. Miller, and Robert C. Atchley, *Families in Later Life* (Belmont, CA: Wadsworth, 1979), 108; see also Nina Nahemow, "The Changing Nature of Grandparenthood," *Medical Aspects of Human Sexuality* 19 (April 1985): 185–90.
5. Andrew J. Cherlin, *Marriage, Divorce, Remarriage* (Cambridge: Harvard University Press, 1981).
6. Norman B. Ryder, "Components of Temporal Variations in American Fertility," in *Demographic Patterns in Developed Societies,* ed. Robert W. Hiorns (London: Taylor and Francis, 1980), 15–54.

7. Susan Cotts Watkins, Jane A. Menken, and John Bongaarts, "Continuities and Changes in the American Family," paper presented at the annual meeting of the Social Science History Association, Toronto, 25–28 October 1984.

8. Bureau of the Census, *Historical Statistics,* ser. A30, A31, A36, and A37; U.S. Bureau of the Census, *Current Population Reports,* ser. P–25, no. 965, "Estimates of the Population of the United States, by Age, Sex, and Race: 1980 to 1984" (Washington, D.C.: U.S. Government Printing Office, 1985), Table 2; and U.S. Bureau of the Census, *Current Population Reports,* ser. P–25, no. 937, "Provisional Projections of the Population of States by Age and Sex: 1980 to 2000" (Washington, D.C.: U.S. Government Printing Office, 1983), Table 2.

9. Bureau of the Census, *Historical Statistics,* ser. R3.

10. Bureau of the Census, *Statistical Abstract,* 558.

11. Bureau of the Census, *Historical Statistics,* ser. Q175.

12. Ibid., ser. Q202–203; and Bureau of the Census, *Statistical Abstract,* 619.

13. Daniel Scott Smith, "Historical Change in the Household Structure of the Elderly in Economically Developed Societies," in *Aging: Stability and Change in the Family,* ed. Robert W. Fogel et al. (New York: Academic Press, 1981), 91–114; and Smith, "Life Course, Norms, and the Family System of Older Americans in 1900," *Journal of Family History* (Fall 1979): 285–98.

14. U.S. Bureau of Labor Statistics, "New Worklife Estimates," bull. 2157 (Washington, D.C.: U.S. Government Printing Office, November 1982).

15. U.S. Social Security Administration, *Social Security Bulletin, Annual Statistical Supplement: 1983* (Washington, DC: U.S. Government Printing Office, 1983), 106.

16. Bureau of the Census, *Historical Statistics,* ser. D803.

17. Ibid., ser. D182 and D196.

18. Bureau of the Census, *Statistical Abstract,* 381 and 484.

19. U.S. Bureau of the Census, *Current Population Reports,* ser. P–60, no. 144, "Characteristics of the Population below the Poverty Level: 1982" (Washington, D.C.: U.S. Government Printing Office, 1984), Table 1.

20. On the differing value of elderly men and women to the household economy, see Michael Anderson, "The Impact on the Family Relationships of the Elderly of Changes since Victorian Times in Governmental Income-Maintenance Provision," in *Family, Bureaucracy, and the Elderly,* ed. Ethel Shanas and Marvin B. Sussman (Durham, NC: Duke University Press, 1977), 36–59; and Virginia Yans-McLaughlin, *Family and Community: Italian Immigrants in Buffalo, 1880–1930* (Ithaca, NY: Cornell University Press, 1977), especially 173 and 257. On intergenerational cooperation in general, see Anderson, *Family Structure in Nineteenth Century Lancashire* (Cambridge: Cambridge University Press, 1971); Tamara K. Hareven, "Historical Changes in the Timing of Family Transitions," in *Aging: Stability and Change in the Family,* ed. Fogel et al. (New York: Academic Press, 1981), 143–65; and Louise A. Tilly and Joan W. Scott, *Women, Work, and Family* (New York: Holt, Rinehart, and Winston, 1978).

21. Tamara K. Hareven, *Family Time and Industrial Time* (Cambridge: Cambridge University Press, 1982).

22. Gunhild O. Hagestad, "Continuity and Connectedness," in *Grandparenthood,* ed. Vern Bengtson and Joan Robertson (Beverly Hills, CA: Sage Publications, 1985), 31–48. Quoted at p. 33.

23. Yans-McLaughlin, *Family and Community,* 256.

24. E. Franklin Frazier, *The Negro Family in the United States,* rev. and abr. ed. (Chicago: University of Chicago Press, 1939), chap. 7.
25. Jane Range and Maris A. Vinovskis, "Images of the Elderly in Popular Magazines: A Content Analysis of *Littell's Living Age,* 1845–1882," *Social Science History* 5 (Spring 1981): 123–70.
26. Anderson, "The Impact on Family Relationships," 58.
27. Ibid., 57.
28. Ibid., 59.
29. A. R. Radcliffe-Brown, "On Joking Relationships," *Africa* 13 (1940): 195–210.
30. Dorian Apple (Sweetzer), "The Social Structure of Grandparenthood," *American Anthropologist* 58 (August 1956): 656–63.
31. Carl N. Degler, *At Odds: Women and the Family in America from the Revolution to the Present* (New York: Oxford University Press, 1980).
32. Daniel Scott Smith, "Parental Control of Marriage Patterns: An Analysis of Historical Trends in Hingham, Massachusetts," *Journal of Marriage and the Family* 35 (August 1973): 419–28.
33. Viviana A. Zelizer, *Pricing the Priceless Child* (New York: Basic Books, 1985), 139.
34. Ibid., 139.
35. Degler, *At Odds,* 14.
36. Ernest W. Burgess and Harvey J. Locke, *The Family: From Institution to Companionship* (New York: American Book Company, 1945), 26–27 and 28.
37. David Hackett Fischer, *Growing Old in America* (New York: Oxford University Press, 1978); and W. Andrew Achenbaum, *Old Age in the New Land* (Baltimore: The Johns Hopkins University Press, 1978).
38. W. Andrew Achenbaum, *Shades of Gray: Old Age, American Values, and Federal Policies since 1920* (Boston: Little, Brown and Co., 1983), 12.
39. Fischer, *Growing Old,* 72.
40. Ibid., 154.
41. Carole Haber, *Beyond Sixty-Five: The Dilemma of Old Age in America's Past* (Cambridge: Cambridge University Press, 1983), 5.
42. William L. O'Neill, *Divorce in the Progressive Era* (New York: Franklin Watts, New Viewpoints, 1973).
43. Arthur Kornhaber and Kenneth L. Woodward, *Grandparents/Grandchildren: The Vital Connection* (New York: Doubleday, Anchor Press, 1981), 240.
44. Ibid., 145.
45. See Hagestad, *Continuity and Connectedness,* for a catalog of these epithets.
46. Ibid.
47. Lillian E. Troll, "Grandparents: The Family Watchdogs," in *Family Relationships in Later Life,* ed. T. Brubaker (Beverly Hills, CA: Sage Publications, 1983), 63–74.
48. Sheppard G. Kellam, Margaret A. Ensminger, and J. T. Turner, "Family Structure and the Mental Health of Children," *Archives of General Psychiatry* 34 (1977): 1012–22.
49. Scott H. Beck and Rubye W. Beck, "The Formation of Extended Households During Middle Age," *Journal of Marriage and the Family* 46 (May 1984): 277–87.
50. William C. Hays and Charles H. Mindel, "Extended Kinship Relations

in Black and White Families," *Journal of Marriage and the Family* 35 (February 1973): 51–57.

51. Cherlin, *Marriage, Divorce, Remarriage.*

Chapter 3 / Styles of Grandparenting

1. Lillian E. Troll, Sheila J. Miller, and Robert C. Atchley, *Families in Later Life* (Belmont, CA: Wadsworth, 1979).

2. See, for example, Bernice Neugarten and Karol Weinstein, "The Changing American Grandparent," *Journal of Marriage and the Family* 26 (May 1964): 199–204; Vivian Wood and Joan Robertson, "The Significance of Grandparenthood," in *Time, Roles, and Self in Old Age,* ed. Jaber Gubrium (New York: Human Sciences Press, 1976), 279–304; and Helen Kivnick, "Grandparenthood: An Overview of Meaning and Mental Health," *The Gerontologist* 22 (1982): 59–66.

3. For example, in the General Social Survey (a national study conducted almost every year between 1972 and 1982 by the National Opinion Research Center) only 13 percent of all respondents described their marriages as "not too happy" (as opposed to "pretty happy" or "very happy"). And only 7 percent described the amount of satisfaction they got from their family lives as "some," "a little," or "none." See James A. Davis, *General Social Surveys, 1972–1982: Cumulative Codebook* (Chicago: National Opinion Research Center, 1982).

Chapter 4 / Grandparental Careers

1. The percentages add to 101 percent due to rounding error.

2. Bernice L. Neugarten and Karol K. Weinstein, "The Changing American Grandparent," *Journal of Marriage and the Family* 26 (May 1964): 199–204.

3. Ibid., 202.

4. See Lillian E. Troll, "Grandparenting," in *Aging in the '80s,* ed. L. Poon (Washington, D.C.: American Psychological Association, 1980), 475–81.

5. Lillian E. Troll, Sheila J. Miller, and Robert C. Atchley, *Families in Later Life* (Belmont, CA: Wadsworth, 1979), 120.

6. Reuben Hill et al., *Family Development in Three Generations* (Cambridge, MA: Schenkman, 1970), 63 and chap. 3 passim.

7. Ibid., table 3.01, 62.

8. Ibid., 69.

9. Louis Harris and associates, *The Myth and Reality of Aging in America* (Washington, D.C.: National Council on the Aging, 1975).

10. Marcel Mauss, *The Gift* (New York: Norton, 1967); originally published in France as *Essai sur le don* in 1925.

11. David M. Schneider, *American Kinship: A Cultural Account,* 2d ed. (Chicago: University of Chicago Press, 1980).

Chapter 5 / Variations

1. See Mirra Komarovsky, *Blue-Collar Marriage* (New York: Random House, Vintage Books, 1967); and Bert N. Adams, *Kinship in an Urban Setting* (Chicago: Markham, 1968).

2. Claude S. Fischer, "The Dispersion of Kinship Ties in Modern Society: Contemporary Data and Historical Speculation," *Journal of Family History* 7 (1982): 353–75.

3. See Reuben Hill et al., *Family Development in Three Generations* (Cambridge, MA: Schenkman, 1970).

4. In these regression analyses, which are not shown in appendix 2, the sample was restricted to families in which the study children were living with both parents. Once a variable measuring the closeness between the grandparent and the child's mother was entered into the equation, a second variable measuring closeness with the father did not add significant predictive power, even for paternal grandparents.

5. These are questions 42a, 42c, and 42d, respectively. A factor analysis showed that these three questions formed a single dimension. A three-item scale formed from the responses to the three questions had an alpha reliability of 0.54.

6. See the multiplicative coefficient for the "family rituals" variable in table A–3 of appendix 2.

7. See table A–3 of appendix 2.

8. The correlation between the log-transformed measures of frequency of visits and frequency of telephone calls over the past twelve months (the two dependent variables analyzed in appendix 2) is 0.49.

9. See the multiplicative coefficient for maternal versus paternal lineage in table A–4 of appendix 2.

10. Sheila R. Klatzky, *Patterns of Contact with Relatives* (Washington, D.C.: American Sociological Association, 1971).

11. See table A–5, column two, "Adjusted Means."

12. See table A–6, column two, "Adjusted Means."

13. For a review of this literature see Lillian E. Troll, Sheila J. Miller, and Robert C. Atchley, *Families in Later Life* (Belmont, CA: Wadsworth, 1979), especially 99–101.

14. For the classic statement, see Talcott Parsons and Robert F. Bales, *Family, Socialization, and the Interaction Process* (New York: The Free Press, 1955).

15. See Gunhild O. Hagestad, "Problems and Promises in the Social Psychology of Intergenerational Relations," in *Aging: Stability and Change in the Family,* ed. Robert W. Fogel et al. (New York: Academic Press, 1981), 11–46; and Lillian E. Troll, "Grandparenting," in *Aging in the '80s,* ed. L. Poon (Washington, D.C.: American Psychological Association, 1980), 475–81.

16. Hagestad, "Problems and Promises."

17. See, for example, Troll, Miller, and Atchley, *Families in Later Life,* 108.

18. A series of log-linear models of the log-odds of scoring high on the exchange scale were estimated, using age of grandparent, sex of grandparent, sex of grandchild, and frequency of visits in the previous twelve months as predictor variables. The inclusion of a parameter representing the interaction of sex of grandparent and sex of grandchild improved the fit of the model significantly.

19. Hagestad, "Problems and Promises."

20. In these and subsequent comparisons, "whites" actually refers to all nonblack grandparents, including a small number of Asian-Americans.

21. See, for example, Carol B. Stack, *All Our Kin* (New York: Harper and Row, 1974).

22. Harriette P. McAdoo, "Factors Related to Stability in Upwardly Mobile Black Families," *Journal of Marriage and the Family* 40 (1978): 761–76.

23. Robert S. Lynd and Helen Merrell Lynd, *Middletown: A Study in American Culture* (New York: Harcourt, Brace and World, 1929); and Robert S. Lynd, *Middletown in Transition: A Study in Cultural Conflicts* (New York: Harcourt, Brace and World, 1937).

24. Theodore Caplow et al., *Middletown Families: Fifty Years of Change and Continuity* (Minneapolis: University of Minnesota Press, 1982), 15.

25. For a review of these trends, see Andrew J. Cherlin, *Marriage, Divorce, Remarriage* (Cambridge: Harvard University Press, 1981).

Chapter 6 / A Special Case: Grandparents and Divorce

1. George S. Rosenberg and Donald F. Anspach, "Sibling Solidarity in the Working Class," *Journal of Marriage and the Family* 35 (February 1973): 108–13; J. W. Spicer and G. D. Hampe, "Kinship Interaction after Divorce," *Journal of Marriage and the Family* 37 (February 1975): 113–19; Constance R. Ahrons and Madonna E. Bowman, "Changes in Family Relationships following Divorce of Adult Child: Grandmother's Perceptions," in *Impact of Divorce on the Extended Family*, ed. Esther Oshiver Fisher (New York: Haworth Press, 1982), 49–68; Gunhild O. Hagestad, Michael A. Smyer, and Karen L. Stierman, "Parent-child Relations in Adulthood: The Impact of Divorce in Middle Age," in *Parenthood: Psychodynamic Perspectives*, ed. R. Cohen, S. Weissman, and B. Cohler (New York: Guilford Press, 1983); Frank F. Furstenberg, Jr., and Graham B. Spanier, *Recycling the Family: Remarriage after Divorce* (Beverly Hills, CA: Sage Publications, 1984); and Sara H. Matthews and Jetse Sprey, "The Impact of Divorce on Grandparenthood: An Exploratory Study," *The Gerontologist* 24 (February 1984): 41–7.

2. Furstenberg and Spanier, *Recycling the Family;* and Colleen L. Johnson, "Grandparenting Options in Divorcing Families," in *Grandparenthood*, ed. Vern L. Bengtson and Joan R. Robertson (Beverly Hills, CA: Sage Publications, 1985).

3. Larry L. Bumpass, "Children and Marital Disruption: A Replication and Update," *Demography* 21 (February 1984): 71–82.

4. Bert N. Adams, *Kinship in an Urban Setting* (Chicago: Markham, 1968); and Claude S. Fischer, "The Dispersion of Kinship Ties in Modern Society: Contemporary Data and Historical Speculation," *Journal of Family History* 7 (Winter 1982): 353–75.

5. Adams, *Kinship in an Urban Setting.*

6. These terms are not meant to have a legal meaning. None of the grandparents actually had legal custody of the study child. Moreover, in about one-ninth of the families the separation had not been followed yet by a divorce. The distinction merely is meant to reflect the difference between being on the side of the parent who was still living with child at the time of the 1981 parent interview and being on the side of the parent who was not.

7. The figures in this section come from responses to questions 91a through 91g.

8. Vern L. Bengtson and Joseph A. Kuypers, "Generational Difference and the Developmental Stake," *Aging and Human Development* 2 (1971): 249–60.

9. We created a two-item scale by combining the responses to the questions about providing financial assistance and co-residing, and then we regressed that scale on a number of determinants of assistance for the custodial grandparents. (The number of noncustodial grandparents was too small to support a similar analysis.)

10. Weighted sample sizes are 111 for the custodial grandparents and 29 for the noncustodial grandparents. A chi-squared test of the difference between the responses of the two groups yielded a value of 24.3, which is statistically significant at the .001 level.

11. Weighted sample sizes are 111 for custodial grandparents and 29 for the noncustodial grandparents. Chi-squared = 4.88, $p < .10$.

12. Grandparents who were living with the study children were excluded. Weighted sample sizes are 100 for the custodial grandparents and 28 for the noncustodial grandparents. Chi-squared = 10.1, $p < .10$.

13. In the 1981 interview, the parent also was asked how many days the nonresident biological parent (typically her ex-husband) had spent with the study child in the previous twelve months. In the entire study, about half of the children had not seen their biological parent living outside the home in the previous twelve months; only about one out of six children had seen their nonresident biological parent on the average of once a week or more. However, when we examined the patterns of contact for the families of the thirty (unweighted) noncustodial grandparents for whom complete information existed, we found much greater amounts of contact between the child and the nonresident parent than for the study as a whole. The study child had not seen the nonresident biological parent during the previous twelve months in only four of the thirty families; in ten out of thirty, the child had seen the nonresident biological parent on the average of once a week or more. Thus, it appears that the parents who were interviewed in 1981 tended to give us the names of their ex-in-laws in cases where there still was contact between their ex-spouse and the child.

14. For the eleven noncustodial grandparents who lived within ten miles of the study child there was a correlation of .38 between the natural logarithm of the number of days the biological parent living outside the home had spent with the study child in the twelve months prior to the 1981 interview and the natural logarithm of the number of visits the grandparent had had with the study child in the twelve months prior to the 1983 interview.

15. For a discussion of kin-keepers, see Gunhild O. Hagestad, "Grandparenthood and Intergenerational Processes: Continuity and Connectedness," in *Grandparenthood,* ed. Bengtson and Robertson.

16. Andrew J. Cherlin, *Marriage, Divorce, Remarriage* (Cambridge: Harvard University Press, 1981).

17. In cases of remarriage, there were more custodial grandparents who reported moderately frequent visits with the study child: 60 percent of those in remarried families had seen the study child at least once every two or three months but not more than once a week during the previous year, compared to 32 percent among the families with no remarriage. But custodial grandparents

in remarried families were less likely to report very frequent contact (more than once a week) or very infrequent contact (less than once every two or three months).

18. Compare the first and third columns of table A–7.
19. Compare the first and second columns of table A–7.
20. Compare the third and fourth columns of table A–7.
21. We regressed scores on the scale of parentlike behavior on sets of dummy variables measuring marital disruption, frequency of contact between grandparent and grandchild, race, age, and family income. The analysis was very similar to table A–5 in appendix 2 except that it was restricted to maternal grandparents. Results are available from the authors on request.
22. See also Furstenberg and Spanier, *Recycling the Family.*
23. Keren Brown Wilson and Michael R. DeShane, "The Legal Rights of Grandparents: A Preliminary Discussion," *The Gerontologist* 22 (February 1982): 67–71.
24. About three-fourths of all divorced persons eventually remarry. See Cherlin, *Marriage, Divorce, Remarriage.*
25. See Furstenberg and Spanier, *Recycling the Family;* and Andrew Cherlin, "Remarriage as an Incomplete Institution," *American Journal of Sociology* 84 (November 1978): 634–50.
26. See Colin Murray Parkes and Joan Stevenson-Hinde, eds., *The Place of Attachment in Human Behavior* (New York: Basic Books, 1982).
27. Furstenberg and Spanier, *Recycling the Family.* For a contrary view based on clinical case studies, see Richard A. Kalish and Emily Visher, "Grandparents of Divorce and Remarriage," in Fisher, *Impact of Divorce,* 127–40.
28. Question 99d. The proportion "agreeing" is the proportion who responded "very much" or "somewhat."
29. The percentages responding that they agreed "very much" or "somewhat" with these four statements (questions 99c, 99b, 99e, and 99a) were 36, 31, 39, and 37 percent, respectively.
30. Furstenberg and Spanier, *Recycling the Family.*
31. See David M. Schneider, *American Kinship: A Cultural Account,* 2d ed. (Chicago: University of Chicago Press, 1980).
32. Colleen L. Johnson reaches a similar conclusion in her articles, "A Cultural Analysis of the Grandmother," *Research on Aging* 5 (December 1983): 547–67; and "Grandparenting Options," in *Grandparenthood,* ed. Bengtson and Robertson.
33. Lillian E. Troll, "Grandparenting," in *Aging in the '80's,* ed. L. Poon (Washington, D.C.: American Psychological Association, 1980), 475–81; and Hagestad, "Continuity and Connectedness."

Chapter 7 / The Influence of Grandparents on Grandchildren

1. E. Franklin Frazier, *The Negro Family in the United States* (Chicago: University of Chicago Press, 1939).
2. See, for example, Robert B. Hill, *Informal Adoption Among Black Families* (Washington, D.C.: National Urban League, 1977).

3. Barbara R. Tinsley and Ross D. Parke, "Grandparents as Support and Socializing Agents," in *Beyond the Dyad,* ed. M. Lewis (New York: Plenum, 1984), 161–93. Quoted at p. 166.

4. In the analyses that follow, the grandparents in our survey will be paired with their grandchildren. Since two or three grandparents were interviewed in a minority of families (see appendix 1), we will in some cases be counting children two or three times in order to preserve the size of the grandparent sample. This procedure gives greater weight to certain families, but it does not, in fact, affect the findings we report.

5. Reuben Hill et al., *Family Development in Three Generations* (Cambridge, MA: Schenkman, 1970).

6. As evidence of this underestimate, we found that children from disrupted homes were not more likely to list a grandparent as an outside parent than those in intact families; whereas, on both of the other indicators of the importance of grandparents, children from divorced families were much more likely to rely on their grandparents.

7. The discussion presented here is based on a multiple classification analysis of the proportion of children scoring one or more on the index. Predictor variables included mother's education, the geographical proximity of the grandparents (whether at least one set lived within an hour's drive), whether the parent received help (such as with childcare, errands, housework, or home repairs) from either set of grandparents, whether the parent received financial assistance, the parents' income, their race, and whether their marriage had been disrupted. Three variables—mother's education, proximity, and help—remained statistically significant predictors even after controls for all other variables. The figures reported in the text are adjusted percentages based on this analysis.

8. See Richard A. Kalish and Ann I. Johnson, "Value Similarities and Differences in Three Generations of Women," *Journal of Marriage and the Family* 34 (February 1972): 49–54; Lillian Troll and Vern Bengtson, "Generations in the Family," in *Contemporary Theories about the Family,* vol. 1, ed. Wesley R. Burr (New York: The Free Press, 1979), 127–61; C. Ray Wingrove and Kathleen F. Steven, "Age Differences and Generational Gaps: College Women and their Mothers' Attitudes toward the Female Roles in Society," *Youth and Society* 13 (1982): 289–301; and Vern L. Bengtson, "Research Across the Generation Gap," in *Relationships: The Marriage and Family Reader,* ed. J. Rosenfeld (Glenview Il.: Scott Foresman, 1982), 50–64. For a review of the literature on intergenerational transmission of values, see Vern L. Bengtson et al., "Generations, Cohorts, and Relations between Age Groups," in *Handbook of Aging and the Social Sciences,* ed. R. H. Binstock and E. Shanas (New York: Van Nostrand Reinhold and Company, 1985), 304–38.

9. We examined each of the potential sources of grandparent-child agreement in turn and discovered that none significantly elevated the mean level of agreement of family values. Specifically, we did not find that an increase on the family ritual scale, the index of exchange, the measure of closeness, or any of the other indicators of strong involvement by the grandparent produced a higher level of agreement in family values between grandparents and their grandchildren. (For a discussion of how agreement was calculated, see appendix 4.)

10. See, for example, Ira Mothner, *Children and Elders: Intergenerational Relations in an Aging Society* (New York: Carnegie Corporation of New York, Aging Society Project, 1985); and Arthur Kornhaber and Kenneth L. Woodward, *Grandparents/Grandchildren: The Vital Connection* (New York: Doubleday, Anchor Press, 1981).

11. Tinsley and Parke, "Grandparents as Support and Socializing Agents."

12. Sheppard G. Kellam, Margaret A. Ensminger, and J. T. Turner, "Family Structure and the Mental Health of Children," *Archives of General Psychiatry* 34 (1977): 1012–22.

13. Frank Furstenberg and Paul A. Allison, "How Divorce Affects Children: Variations by Age and Sex," paper presented at the Society for Research on Child Development, Toronto, April 1985.

14. See references at note 8.

Chapter 8 / The Future of Grandparenthood

1. Joseph Veroff, Elizabeth Douvan, and Richard A. Kulka, *The Inner American: A Self Portrait from 1957 to 1976* (New York: Basic Books, 1981), 535. For a similar argument about the changing basis of self-identity in American society, see Ralph H. Turner, "The Real Self: From Institution to Impulse," *American Journal of Sociology* 81 (March 1976): 989–1016.

2. Ronald Freedman, Ming-Cheng Chang, and Te-Hsiung Sun, "Household Composition, Extended Kinship, and Reproduction in Taiwan," *Population Studies* 36 (1980): 395–411.

3. Robert N. Bellah et al., *Habits of the Heart: Individualism and Commitment in American Life* (Berkeley: University of California Press, 1985), 90.

4. See the discussion in chapter 2 of research by Dorian Apple Sweetzer and A. R. Radcliffe-Brown.

5. Arthur Kornhaber, "Grandparenthood and the 'New Social Contract'," in *Grandparenthood,* ed. Vern L. Bengtson and Joan F. Robertson (Beverly Hills, CA: Sage Publications, 1985), 159–71. Quoted at p. 159.

6. Arthur Kornhaber and Kenneth L. Woodward, *Grandparents/Grandchildren: The Vital Connection* (New York: Doubleday, Anchor Press, 1981), 147.

7. For an elaboration of the importance for marriage and fertility of "wealth flows" to and from the older generation, see John C. Caldwell, *Theory of Fertility Decline* (London: Academic Press, 1982).

8. Samuel H. Preston, "Children and the Elderly in the U.S.," *Scientific American* 251 (December 1984): 44–49.

9. Greg J. Duncan, Martha Hill, and Willard Rodgers, "The Changing Economic Status of the Young and Old," paper presented at the Workshop on Demographic Change and the Well-Being of Dependents, National Academy of Sciences, 5–7 September 1985.

10. Charles F. Westoff, "Marriage and Fertility in the Developed Countries," *Scientific American* 239 (December 1978): 51–57.

INDEX

Index

Index

Index